Introduction

Driving into St Augustine, a coastal town in north-eastern Florida, I'm bombarded by signs pointing me to the Fountain of Youth. Like the millions of visitors over the last century that have made a pilgrimage to this place, the oldest attraction in the longest continuously inhabited European-established settlement in what is now the United States (eclipsing in popularity the historic city centre, the Bluebird of Happiness statue and the National Shrine of Our Lady of La Leche by several hundred visitors per day), I'm here because of the slimmest chance that I might regain a little bit of vim and vigour in my ageing body. I've hit the fifty-year mark and though I've never believed it would happen to me after a lifetime of athletics and moderately clean living, my eyes appear to be going, and all the sports injuries I've played through or endured are reminding me they're there. I'm a psychologist with decades of scientific training behind me, and so I'd like to think that I'd rely on empirical evidence if I chose to treat my 'age' in any way, but today I've convinced myself it's worth the mosquito bites and the humidity and $22.95 for an adult entry ticket and the tiniest plastic cup of water: I'm going to get a taste of eternal life.

The story goes that one day in 1901, Dr Luella Day McConnell rode into St Augustine wrapped in ermine with a diamond in her tooth and bought up a nice piece of land using the money she'd made prospecting for gold in the Yukon Territory in Canada. She was already notorious; rumour had it that she'd abandoned two husbands and her medical practice in Chicago, that the British government had tried to assassinate her, and that she'd made a dramatic return from the dead. The locals dubbed her Diamond Lil.

Lil's new property had two things going for it: ruins of a couple hundred years of Spanish settlements and 3,500 years of Native American history, and a small spring.[1] She fixated on the spring, which she fiercely marketed as the spot where the conquistador Juan Ponce de León landed in 1513 (it wasn't) while in search of the fountain of youth (he wasn't). For a good twenty years, she spread remarkable stories about the spring water, until her untimely death in 1927 – at the wheel of a car when it met with the bottom of a ditch – which is still rumoured by some to have been a murder. After she died, the property and its popular attraction was sold to Walter B. Fraser for $100,000 – that's almost $2 million in today's money – who maintained it as an attraction to educate the public about Ponce de León,[2] and to exploit the spring for its life-giving properties.

Full disclosure: not one person who has visited Lil's spot has (so far) lived beyond the longest-recorded human lifespan (122 years and 164 days), and the vast majority have in fact since died – due to circumstances, of course, entirely unrelated to the fountain. And I know this uncompromisingly expensive thimbleful of Department of Health-inspected water won't do anything about my creaking back, but my ceremonial sip is part of an immortality ritual that stretches back to the beginning of time, and which has kept even the most rational people in history captivated. The possibility of keeping death at bay and staving off the ravages of age makes us feel good. It gives us the illusion that we have control over nature. People throughout history have chased the dream: popes, emperors, kings, and everyday folks too. And it has a new set of explorers in

The Immortalists

ALSO BY ALEKS KROTOSKI

Untangling the Web

The Immortalists

The Death of Death and the Race for Eternal Life

ALEKS KROTOSKI

THE BODLEY HEAD
LONDON

1 3 5 7 9 10 8 6 4 2

The Bodley Head, an imprint of Vintage, is part of the
Penguin Random House group of companies

Vintage, Penguin Random House UK, One Embassy Gardens,
8 Viaduct Gardens, London SW11 7BW

penguin.co.uk/vintage
global.penguinrandomhouse.com

First published by The Bodley Head in 2025

Typeset in 11.9/16pt Sabon Next LT Pro by Six Red Marbles UK, Thetford, Norfolk
Printed and bound in Great Britain by Clays Ltd, Elcograf S.p.A.

The authorised representative in the EEA is Penguin Random House Ireland,
Morrison Chambers, 32 Nassau Street, Dublin D02 YH68

A CIP catalogue record for this book is available from the British Library

HB ISBN 9781847928504
TPB ISBN 9781847928511

Penguin Random House is committed to a sustainable future
for our business, our readers and our planet. This book is made
from Forest Stewardship Council® certified paper.

To the Grim Reaper: congratulations on your retirement.

Contents

PART V
Longevity Nation

the twenty-first century: technologists who struck it rich in the Silicon Valley gold mines and who believe that they, with their unique skills and the technology at their disposal, can keep death away.

∞

I have been a technology reporter for twenty-five years, chronicling the rise of the web from its early days through booms and busts. I fell into early Bulletin Board Systems (BBSs) when there was no World Wide Web, and got sucked into the internet's Usenet forums around the time the very first spam message was sent out. My closest friends were the people who created the first BBC webpages, and hosted the first open Wi-Fi hotspots, and designed the Wi-Fi symbol. I was one of the first thousand people to use both Twitter and Instagram. But I wasn't a maker; I was documenting the breaking news from this other, virtual world. Later, after my curiosity in the subject became more consuming, I studied it for my MSc and my PhD, and for fellowships at the University of Oxford and the London School of Economics. I have continued to report on it, investigating the at-times dangerous ways the online world outpaces the old system. And now I also have the honour of provoking the next crop of designers to think about the holistic human using their machines, in my lectures at NYU's ITP MA, the information technology programme in the Tisch School of the Arts.

Over the years, I have become impervious to the hype coming out of Silicon Valley. Instead I use my academic chops to track the cultures that built the digital world, and the people who have gathered there. I have railed, frustrated, against developers who are often entitled, overzealous, and quick to reduce the complexities of human psychology into the simplistic solutions that work – for a while – until the shortcomings of their disruptive innovations inevitably collide with society at large.

I love technology and I love technologists, but my job has been to call them out on their impossible promises and to show them how their extraordinary inventions don't just help us, but also – because of how they're designed – can harm us too. For the last quarter century, I've written, broadcast and reported for the *Guardian*, BBC TV and BBC radio – for the award-winning *Digital Human* series and, since 2024, *The Artificial Human* – and many other outlets. I've written books about how the internet doesn't really change who we are as people, just how we do what we do. My motivation has never been to ban technology outright, but to educate users to recognise that we can refuse what tech wants us to do, and we should criticise it like we would other forms of media, or public policy.

Very often, I am asked by concerned friends, family and even strangers whether it's possible to stop being 'corrupted' by tech, or 'addicted' to it. So many people feel overwhelmed by what seems like an onslaught of change perpetuated by the technology industry – so many disruptions, all the time. And now, on the cusp of the AI age, once again we are told everything is going to change.

The developers, the funders, the Silicon Valley machine are preaching a total upheaval of humanity. They say that their tools must be prioritised above all. And because they feel they've already disrupted everything else, now they believe they have the power to disrupt death.

In these pages, you won't hear about the supplements I take or the health routines I follow. This is not a lifestyle guide. What you will discover is cults, blind faith, and the pursuit of absolute power. You'll hear about remarkable treatments that mostly don't work, some that we don't know why they do, and some where the jury's still out. You'll get to peek into the motivations of the immortalists who are restructuring our world towards their vision of utopia: what they believe, why they believe it, and where this rubs up dangerously against ethics and decency. And at the end of it, you get to make your own judgement call about whether you think their version of tomorrow is the one you want. If so, enjoy the ride. If not, this book will give you what you need to know to fight back.

In Part 1, we'll investigate how the lifeblood of Silicon Valley – data – is turning our bodies into machines. We'll see how this has been translated into lifestyle hacks and medical innovations, and how this has extended our healthy lives, but how it's also caused us to lose touch with ourselves.

In Part 2, we are introduced to the true believers: the inves- ✔ tors, the scientists and the fanatics who believe that we truly are on the cusp of the next evolutionary step in humanity's journey, and that they are the pioneers, or the lab rats, who will break through biological limits in order to prove that it's possible to have a radically extended life. We also meet the entrepreneurs who are exploiting this passion, and discover the structures that are in place to keep all of us safe.

In Part 3, we'll find out how a technological mindset takes ✗ this further, into futuristic scenarios in which immortal life means outrunning time, uploading our brains to computers, and living forever in computer server farms on Jupiter. We'll meet the billionaires taking steps to make these wild ideas a reality, and the scientists they are investing in.

In Part 4, we take a breath, and ask what all this extra life might mean for how we live it. The future, as science-fiction author William Gibson wrote, is not evenly distributed, so who benefits from these extraordinary technologies, who's left behind, and – most importantly – who decides who gets to live forever?

In Part 5, we answer that question. We look at the context of the political landscape of 2025, as Silicon Valley gains unprecedented access to the highest positions of power in the western world, and restructures regulatory systems from within, so its vision of the future can come to pass. We also follow the grassroots immortality movements created by those who don't have the ears of the powerful, but believe in the sermons of the tech billionaires, and are building a long-lived tomorrow, today.

This book began in a different form, as a chart-topping radio and podcast investigation called *Intrigue: The Immortals* for BBC Radio 4, broadcast in 2023. For that series, I travelled around the world to speak with leaders in ageing science, biotechnology and Silicon Valley investment. I heard stories from people who believed in immortality, and people who said they didn't (but really wanted to). I built the foundations of the ideas you will read in this book for that project.

But over the last few years, I've dug into the history books, the policy papers and the academic journals, and I've attended lectures, visited archives, and spoken with many more people than appeared in the series. I have expanded on the themes and the stories, going much deeper into the motivations, ethical conundrums and doctrines that drive the belief that right now is the moment when we are on the cusp of immortality, and that these immortalists are the people who will give it to us.

What do I mean when I talk about 'immortalists'? Well, there are several different kinds of immortality in this book. There's the literal live-forever kind, which is dominated by the people I have described as techno-fundamentalists since I first started talking about this group of tech believers in 2012: highly intelligent, mathematically minded computer scientists, philosophers and hopefuls who have such unwavering faith in technology they believe beyond a shadow of a doubt that it will be the chalice of everlasting life. This is where traditional religion finds its way into the story – people in this community believe in a higher power: technology, and its archangel rationality. They imagine that, by the grace of technological progress, they will merge with artificial intelligence and become post-human, and will live forever in a state of surpassing bliss and delight. This dogma is inspiring what some of the people I have spoken with for this investigation describe as 'moral catastrophe.'

Another version of immortality also comes out of the techno-fundamentalism camp, but is less apocalyptic. They also believe in living forever, but that there won't be a big bang. Instead, we will experience an ever-evolving partnership between humans and technology, through which our moral failings will be solved, piecemeal, one tech innovation at a time. Life will continue not by merging with tech, but by using it to slow down the onslaught of time, thereby allowing medical science to develop treatments that will heal us. Medical science is trying to keep up with them, but it's mostly outpaced by the greed that the open market is generating, which is causing a lot of friction. These immortalists believe they are living proof of age reversal, physical rejuvenation, and stopping time. They are the researchers at the fringes of 'acceptable' science, and the biohackers who are using their findings to optimise themselves with data. But, as I will argue, lifestyle disciples have lost touch with what it means to feel 'well' – and this is what's driven the marketplace for supplements and unproven treatments.

And then there's the kind of immortalists who I believe are the most dangerous of all. These are the people who are hungry for the same status that Diamond Lil achieved, by living forever through their legacies. They seek to reconstruct the infrastructure of the world we live in so that they can live forever. Already, and in plain sight, they are restructuring sovereign nation states, pushing forward an agenda of technological acceleration at any cost, and making deals with people in the highest political offices. Along with their nation-state-building activities, they are also pouring money into schemes that give them first dibs at unimaginably long lives. These powerful people aren't searching for the fountain of youth; they're building it, and their first attempts are revealing: their solutions are ableist, classist and racist. Some have even described them as eugenicist.

Philosopher Stephen Cave, director of the Leverhulme Centre for the Future of Intelligence at the University of Cambridge, wrote in his 2012 book *Immortality: The Quest to Live Forever and How It Drives Civilization* that our 'will to immortality' has created the engine of innovation, in part because we have a determination to survive and extend into the future (which Cave argues is something we share with all life forms), but also because we know we are 'cursed, not only to die, but to know that we must'.[3] In the past, we've seen it in religious movements, millennia of philosophy, the rise of cities, the evolution of science and technology, and the creative output of the arts. This book is about tomorrow's civilisation, brought about by the solutions to the human condition that technologists are creating today.

Absolutely central to this is how the computer scientists and mathematicians in Silicon Valley define 'human', which is why the first chapter breaks this down in detail. What do they imagine are the building blocks of humanity? I urge you to compare how this fits into *your* idea of what makes us 'human'. For many experts – from biologists to psychologists to religious scholars to philosophers – the way technology is trying to reconstruct our complexity in 1s and 0s is woefully incomplete.

Take an example of this in practice: in early 2025, in an attempt to make it more efficient, Elon Musk – technologist, investor, techno-optimist, and head of the newly created Department of Government Efficiency (DOGE) – eviscerated the internal workings of the US federal government by disaggregating the different agencies and branches and viewing them as isolated entities, simplifying the complexity of their national and international activities, and synthesising a solution by directing the remaining employees to use AI. This is the same mindset that is trying to hack ageing.

At the heart of this book is the question about why we avoid death at all. There are people who struggle with thanatophobia, the intense fear of death. Is this that? Or is it actually a generalised fear of an absence of one's own being? Is it, to use today's language, an existential threat? In his 1927 work *Being and Time*, philosopher Martin Heidegger described death as an existential phenomenon, explaining that, as humans, it is impossible to imagine what's through the veil. What we like about living is what we do when we are alive. Becoming dead, he posited, means losing something (life). As American philosopher Thomas Nagel wrote, 'If we are to make sense of the view that to die is bad, it must be on the ground that life is a good and death is the corresponding deprivation or loss, bad not because of any positive features but because of the desirability of what it removes.'[4] Is it possible to eradicate it entirely so we don't have to consider the nature of our being? Or is it too much to deal with so we don't, just as we don't with other concepts classified as 'existential': climate change, nuclear war, and artificial superintelligence?

∞

But before we begin, let's consider a parable.

'Once upon a time, the planet was tyrannized by a giant dragon,' wrote philosopher Nick Bostrom in his 2005 essay 'The Fable of the Dragon-Tyrant'.[5] In this story, a village is terrorised by a terrible monster that lives atop a nearby mountain.

> It demanded from humankind a blood-curdling tribute: to satisfy its enormous appetite, ten thousand men and women had to be delivered every evening at the onset of dark to the foot of the mountain where the dragon-tyrant lived. Sometimes the dragon would devour these unfortunate souls upon arrival; sometimes again it would lock them up in the mountain where they would wither away for months or years before eventually being consumed.

The story of this dragon, the devastation it created and the ultimate triumph of technology has become the immortalists' dominant cultural narrative.

Decades before he imagined the mountain and the dragon, Bostrom had been a solitary and bookish child. 'In Sweden, where I grew up, I knew nobody who was interested in how future technologies might radically change the human condition,' he told me in 2023. It wasn't until after he moved to London in 1996 that he discovered the internet, and it was there, online, that he was finally able to talk freely and at length about his intuition that technology could make us immortal. This seemed 'much more profound than exactly where the border is between two countries, or what the tax rate is, he explained. The people he found online didn't sneer or run away; in fact, they were a like-minded community who encouraged him to imagine how technology could – and should – be used to enhance human capabilities. This group, captivated by the *opposite* of entropy, which is the process of a system's gradual decline into disorder, called themselves Extropians and became a formative collective of philosophers, academics, technologists and science-fiction authors whose ideas about life infected an entire generation of out-there thinkers.

On their online 'listserv' mailing list, Bostrom and tech celebrities such as Ray Kurzweil, inventor and author of *The Age of Intelligent Machines*, grappled with how technology could extend human life ad infinitum. This was more than just an intellectual enterprise: technology was their magic wand, and since the 2000s the Extropians have truly believed that we are in a transitory phase towards a greater intelligence, always driven by the idea of perpetual progress towards wisdom, an indefinite lifespan, and self-actualisation, by removing political, biological and cultural limits. Philosophies splintered from it – such as a modern version of a 1950s thought experiment called transhumanism, the belief shared by billionaires such as Musk, Peter Thiel, Meta's Mark Zuckerberg and OpenAI's Sam Altman that it is a moral imperative to use technology to enhance the body and mind towards immortality.[6]

This is not a community based on vibes – or so they say. Behind this movement there continues to be a doctrine of rational thinking and reasoned evidence; flights of fancy are only acceptable if they can be proven to be probable. But this movement is also based on the faith that we *can* be immortal and, most importantly, on the faith in technology to get us there. This community believes that right now we are on the cusp of the most revolutionary moment in human history: the death of death.

Back to 'The Fable'. Over time, the dragon sacrifices became so commonplace that society bent to its will. The king built more railway tracks to the base of the mountain to make the delivery of mortal souls more efficient. Religious groups emerged to explain what lay on the other side. Scientists developed ways to make the inevitability easier. In short, Bostrom argued, the villagers – we – were being indoctrinated into a 'deathist' culture that counselled 'passive acceptance'. 'When I was writing this, there was in academic bioethics a weird neglect of the, as it were, pro side for regarding ageing as a problem to be addressed', he told me. 'At that time, there were a whole bunch of different people who had been arguing that actually ageing and death are good. Like it's somehow part of the natural order. And there were all these attempts to find some clever justification for why, even though it seems like a horrible thing, nevertheless we should welcome it or embrace it or accept it.' The Extropians do not.

In his story, the voice of Extropianism is a clever sage known for his radical ideas, who says it should be possible to break out of the cult of mortality, if only there was money to invent the technology to do it. He and the others who agree are laughed out of town. Society says death is the accepted endpoint, and there is no way around it. Except, Bostrom told me, 'I think, with our advancing technical capabilities, it is going to cease to become an inevitability and something that we can prevent if we choose to do so.

'It seems kind of sad that just as we are, you know, developing some maturity, and experience of life, it all starts to get taken away', he said. 'Once you've learned enough to know how to live, then that's about when life comes to an end and you don't really have a chance to use all of that.'

The solution is always more investment, more development, more technology. In the parable, there's a groundswell of public demand for the king to stop everything and pour money into this dragon problem. This is exactly what many immortalists are ✗ trying to generate today. They expect that eventually, after generations of social and philosophical debates about the morality of dying and the purpose of life, today's scientists, like the scientists in the fable, will develop a material that penetrates the dragon's scales, and the kingdom will be saved. Technology and the people behind it are the heroes of Bostrom's work: 'The great wheel of invention' will always continue to accelerate – as it does in 'Dragon-Tyrant'.

Technology really is accelerating scientific discoveries – extraordinary findings that are ushering in a new way of thinking about and treating the underlying biological mechanisms of ageing. Generative AI really is solving decades-old fundamental biochemical problems, and winning technologists Nobel Prizes in Chemistry – science they are not trained in. But the tools they've invented are cracking biochemical codes: what else can they achieve? Can they really unravel mortal mysteries in our lifetimes? Do we have to die?

We are on the edge of an artificially intelligent future. It's AI ✗ that immortalists believe is going to give us everlasting life. AI, says Bostrom, Musk, Zuckerberg, Altman and other powerful venture capitalists such as Marc Andreessen and Thiel, is what will help us pierce the dragon's scales. Our belief that AI can solve everything is overshadowing the issues we collectively face: overpopulation, environmental catastrophe, inequality, marginalisation, ageism. While their technologies could be directed towards addressing these challenges, the immortalists seem uninterested in doing so – because AI will solve these things too.

The Valley was founded on the principles of innovation, experimentation and learning. For more than seventy-five years, it has moved markets, transformed the workplace, and created prophets. It was born in the halls of academia and industry, from the pocketbooks of government. And thanks to a confluence of smarts, cash, a permissive regulation system and a rebellious counter-culture, it's become the number-one place in the western world to take risks, iterate quickly, and bounce back from failure.

This has empowered technologists to be the modern-day alchemists, thinking they can promise to bring us everlasting life. They have consolidated power in the centres of finance, politics and science. They have become the strategic minds that dictate the direction of travel. And their pursuit of eternal life is driven in large part by their unwavering faith that technology is the thing that will save us.

Maybe they're right. Maybe they will crack the code of life. But here's an alternative ending: maybe it will be humanity that saves the day. I invite you to raise a glass of water to Diamond Lil and start by asking, what are humans anyway?

PART I
Bug Fix

CHAPTER I

Engineer's Syndrome

We are buried beneath the weight of information, which is being ✗
confused with knowledge.

Tom Waits[1]

In January 1996, there were 100,000 websites on the World Wide
Web. Much like human DNA, this new 'information superhigh-
way' seemed to be a wild cacophony of disconnected material, with
no real way to navigate it. There were search engines, but none
were particularly good at delivering what seemed to be the right
answer – until two Stanford University students, Larry Page and
Sergey Brin, came up with their solution. They created a system
that ordered the value of online information using an algorithm
they called PageRank, cheekily named after Larry. More quickly
and effectively than any other technology out there, PageRank
indexed and sorted every destination on the rapidly expanding
web, and used the results to determine the quality of a site.

Google quickly became the search leader, and in thirty short
years the company has revolutionised how we look for information, how we measure success, how we advertise, and how
we learn. But its true worth is much more than that: Google's
value is that we use it to find something we want, whether that's
your pharmacy's opening hours, or what to do on your dream

holiday. Unexpectedly, Page and Brin had invented something that psychologists like myself have been trying to access for almost a century: a window into our desires.

For years, people have been using Google as a kind of subconscious oracle, directly asking it some pretty delicate (or indelicate) questions, or a bunch of questions that add up to a delicate (or indelicate) insight. That we feel comfortable 'confessing' to the internet is similar to how some people might divulge secrets to a priest behind the screen; it feels less consequential than speaking face to face, like our words disappear the moment they are sent into the digital ether. But behind the scenes, tech companies hold on to your internet whims, and use them to make assumptions about you and the world at their discretion.*

By 2008, Google was the biggest compiler of data in the world, reliant on technology being able to process an ever-increasing corpus, in order to index 172 million websites – up 50 million from the year before.[2] What an opportunity, thought Larry Brilliant, who had been the executive director of Google's charitable arm since 2005. Brilliant had announced his greatest wish to an audience at TED in 2006: 'to help build a global system – an early-warning system – to protect us against humanity's worst nightmares.'[3] He and several engineers decided to experiment with the search engine's powerful predictive tool to look for ebbs and flows in the world of public health.

They imagined that if they monitored certain search keywords related to health in real time across Google's enormous database, they might be able to reveal where people are about to catch the flu. They started their experiment in the US. Except for paediatric cases, the flu isn't a required reportable illness in

---

* Of course, what they can actually do is dictated by regulations in each jurisdiction, but most legal frameworks still have yet to catch up with what's actually possible and so are more reactive than proactive in their data privacy statutes.

all US states; the Centers for Disease Control and Prevention (CDC) doesn't have a complete dataset and therefore may not know where to target treatments or interventions.[4] Just under half of all US internet users in 2008 used a search engine;[5] 69.5 per cent of them used Google.[6] So Brilliant and his team wondered if there was a way to close this gap and help save some of the thousands of lives lost every year to influenza by addressing the spread early.

That year, the company released Google Flu Trends (GFT).

GFT started out by tracking a list of automatically selected keywords – from 'flu symptoms' to 'cold/flu remedy' – across the 50 million most-common search queries the site received each week. It cross-referenced those against data reported by the healthcare providers about how many people were going to see their doctor with flu-like symptoms, and found a relationship between the two. The team then looked to see whether the search behaviour might predict a flu outbreak faster than the CDC. Their data suggested they could, up to ten days faster, according to a paper they published in *Nature* in 2009.[7] It seemed that Google technology could turn individual searches into an epidemic alarm bell. Governments were intrigued, and quickly GFT moved beyond the US and spread to more than twenty-five countries.

But soon the shortfalls of Big Data started to appear. After a few years, GFT was predicting twice as many flu-like visits to doctors as the CDC was – which relied on reports from lab testing for the flu. The Google data was massively over-predicting.

As researchers dug into this, they found that the algorithms and machine learning that made the tech project work were at fault, as they assumed people would use certain search terms together when they felt poorly. As one commentator put it, 'the initial version of GFT was part flu detector, part winter detector.'[8] Even after the engineers found the problem, the updated system consistently over-reported, and the predictions were never more accurate than those of the CDC.

This didn't mean that the technology had failed, or that Google should throw the whole thing out. In fact, GFT combined with the CDC's data predicted outbreaks of flu better than either system on its own. This proved to the company, and to the world, that the search giant not only had a dataset of enormous social and public benefit, but that they also had the talent inside to explain and predict. The issue was that they had put too much of their faith in data alone. People, they discovered, are more complicated than computers.

This is a problem the immortalists intend to fix.

The best description I've read about the history of Silicon Valley is that it is a series of audacious claims that have been successfully achieved. Again and again, geeks have shocked us out of our bubbles, pushing us against our boundaries, and forcing us to rethink what we know about what it means to be human. They've done this with how we communicate, how we exchange, what we believe is true, and who we think we are. And what is most extraordinary is that they've done all this by turning us into data.

From my reporting over the last twenty-five years, I've found that there are three dominant paradigms that shape how engineers codify human beings and try to serve us their solutions. The first is the almost universal idea that the human is a machine. This idea has a deep history, but comes with a twist: today, our machines are made of data.

Let's go back to 2014 when I was living in Florence, Italy. My husband and I used to say our flat was around the corner from the Renaissance, and the signs of it were indeed everywhere. A few streets away, in the piazza of Santo Spirito church, was a fountain. Today, it's mostly obscured by the tables and chairs of cafés and restaurants; kids play nearby, lovers perch on its edge and hold hands, and tourists take photos. But five and a half centuries ago, in 1492, it's where a teenage Michelangelo would go at dawn to wash after sneaking out of the monastery. He would be covered in blood and bodily secretions, having spent the night dissecting cadavers in its hospital in order to try to better understand the muscles, tendons and skeleton that are our physiological supporting structures.[9]

It was frowned upon in the Catholic Church to dissect bodies, unless you were a physician, yet across the Arno river was Leonardo da Vinci's studio, where he too practised the art of anatomy. It's also where the artist and inventor dreamed up wild engineering projects. His notebooks overflowed with his models of the world: meticulously designed flying machines, trebuchets, communication devices, and other fantastical objects. One of his most famous images, the *Vitruvian Man*, is just one example of his fascination with the human form, which he described as 'the greatest instrument in nature'. Between 1506 and 1513, his notes also describe thirty human dissections. The first was that of a 100-year-old man whose death he had witnessed only moments before. As Leonardo dissected, he designed a humanoid machine – we'd call it a robot today – that had ropes and pulleys instead of joints and muscles. But he wasn't just interested in how the body was constructed, or intrigued by how it worked; he also was looking to nail down something more ambiguous and less mechanical: the source of our emotions. He never found it. (He did, however, discover that the centre of our circulatory system is the heart, not the liver as previously believed.)[10]

Renaissance artists reproduced the human body at its most

mechanical, and this led to some of the greatest sculpture and art in history. But the cost was defiling our corpses – holy objects that were fearfully and wonderfully made, and, according to the values that were most widely held at that time, the locus of the spirit.

The dissections were sacrilegious deeds: it was against God to imagine that the body could be divided into its functioning parts. 'Prior to the mid-1600s, people had all kinds of ideas about the body having mysterious forces and having been formed by divine interventions,' explained Professor Randolph Nesse, the founding director of the Center for Evolution and Medicine at Arizona State University, when I interviewed him in 2015. 'With Descartes, people started thinking about the body as if it was a machine.'

René Descartes: scientist, devout Catholic, mathematician, and inventor. He made his own models of human bodies less than a hundred years later than Leonardo, and, like the artist, Descartes was fascinated by automata. He believed these 'self-moving things' could illuminate the universal mechanistic principles of the natural world: from recreations of inanimate objects to plants and other animals. He started out building replica mills and clocks, and wondered if their inner workings were like how animals, plants and humans operate. He began to consider what could be purely mechanised, and so reduced to mathematical function. In tandem with his philosophical musings, he believed that applying maths and pure logic to the conundrum of human-ness would reveal a fundamental truth.

But his mechanical models were only ever physical representations of living things. What was the difference between the maquette of nuts and bolts he put together in his workshop, and the things that were alive – particularly humans, made in the image of God?[11] We surely had a divine spark that positioned us at the top of the hierarchy of creatures on Earth. Was it possible

to recreate our emotional and subjective experience in mechanised parts? Descartes needed to find a home for the soul, but he
ended up as stumped as Leonardo: where exactly was the spirit?

He compromised. He split us into mechanical bodies and
immaterial minds. With only the technology available to him
during his time, the latter could not be built. It occupied a space
somewhere distinct from the physical world.

This concept, dualism, gave him the freedom to decon-
struct and master the physical self, using metal parts. He had
the spiritual permission to tinker with the body as one would
tinker with a device. This was the first step towards objectivity,
a basic idea in philosophy usually paired with its contrasting
concept, subjectivity, to distinguish things that can be observed
and measured – e.g. the temperature – and those which are
perceived – e.g. warmth. However, the 'Cartesian split', or the
mind–body problem, had a practical implication too: if we can
know our constituent parts it is possible to figure out more
effective ways of fixing ourselves when we break.

The belief that the body is a machine has become entrenched
in our psyche. 'It's so pervasive,' Nesse says, 'no one even thinks
of it as a metaphor.' And over the centuries, we have continued
to fill in blanks as we have exchanged nuts and bolts for accel-
erometers and electric pulses and 3D-printed valves, developing
ever more precise technologies that can recreate our bodies. (Of
course, this is much easier to do if we only have to deal with the
body. The idea of psychology being used to 'heal' the mind is a
more complex concept.)

Technological metaphors didn't leap directly from mechanical automata to digital code; they updated with each innovation. Steam power gave scientists the language to describe the
body's homeostatic systems: the body seeks equilibrium, so it
releases energy, and circulates it in a particular way. The internal

combustion engine explained how energy is transformed from one state to another; electric power, how our cells have potential. The telephone network helped us to understand how signals travel to the brain. Now, the biological metaphor has become information: data, measurements and taxonomies that can be fixed with the introduction of more code.

'It's quite weird where we are at now, because our machines have changed,' reflects Nesse. 'Our machines are becoming more alive, and we're increasingly becoming more mechanical. We don't think of machines in mechanistic terms anymore ... and our relationship to this technological world has really shifted. It's become a lot more mental or even spiritual. A lot more intangible.' The technology – or, rather, the people who create it – has become far more capable of behaving like a god.

Theoretically, you can break down the anatomy and function of a biological mechanism such as the liver into its constituent parts, identify what appear to be the most essential components, re-engineer those in computer code, and establish a plug-and-play solution to a problem. Like automata, the body can be disaggregated and transmogrified into simple computer code and added to a database. So far, the mind is too complicated to pin down.

But there's a downside to disaggregating and transmogrifying: over-simplification. Combine this with a 'disease' known as Engineer's Syndrome – when technologists apply engineering strategies to a complicated problem in an unrelated field they know little about – and you have a recipe for things going wrong.[12] I remind my IT design students at NYU, when they are coming up with a solution for what they see as a simple fix, that whatever complicated problem they're trying to solve has a context that can't be lost. For example, how can you build a machine that guarantees the user will experience serendipity when all the machine can do is deliver information? It's up to the interpretation of the user to see this information as relevant and valuable, and that they can physically, socially and culturally be able to do so at the moment they're using the device. If the context is lost, all they're solving is a small part of the whole (giving information) – and that's not going to get them top marks.

This syndrome is what's happening with the group of engineers hellbent on 'solving' death. They believe that we are machines that stop functioning, and that all that needs to be done is to fix what's broken using a technological solution which will lead to long – even eternal – life. This profound and complicated problem has become a game of numbers and mathematical equations and a sketchy understanding of the complexity of what makes us alive. This model gives the *body* primacy – the material that can be measured and contained.

Even in 2008, during the period of giddy optimism in the early years of Big Data and GFT, we saw that the power to interpret the immaterial was off the mark. This data-driven approach is one that scientists who are experts in the human body find baffling. In his 2024 book *Why We Die*, Nobel laureate Venki Ramakrishnan describes 'the characteristic arrogance that many physicists and computer scientists display toward biologists' that causes the engineers to miss something crucial. That something may by physical, but it's more likely to be impossible to break down into data, as doctor and prize-winning author of *The Way We Die Now* (2016) Seamus O'Mahoney found when he went to a longevity conference in 2025:

> They are interested only in the biomolecular and the monetizable; I heard a great deal over the four days about AI-designed drugs, glycans, the transcriptome ageing clock, and every other imaginable -ome, but almost nothing on the

complexity of death systems and the social determinants of death and dying. They seemed strangely uncurious about the enemy they have declared war on. Ageing to them is simply a technical problem that can, and will, be fixed.[13]

The idea that it should be possible to map the human body reinforces this attitude. There are currently initiatives from the field of ageing research trying to do this, but they are atomised and compartmentalised. The Human Genome Project began mapping our DNA in 1990, delivering a complete sequence of the human genome in 2003. The Human Brain Project, which began in 2013 and concluded in 2023, published more than 3,000 papers and, among a flood of other research, created atlases for human, rat and macaque brains. In 2024, FlyWire, a consortium of researchers part-funded by the US National Institutes of Health (NIH), released a map of all 140,000 neuronal connections in a fruit fly's brain.[14] But these maps only represent territories we know about. When it comes to how genes affect molecules or traits, or how our neurons affect intuition, the maps are as useful as their medieval counterparts: 'Thar Be Dragons.'

Still, ever optimistic, Valley folks believe that it can be done. 'I think slowly over time we're gonna figure out things here and there to be able to repair ourselves,' Sonia Arrison, an investor in several biotech start-ups, tells me. Arrison is a former policy advisor at the libertarian think tank Pacific Research Institute and founder of 100 Plus Capital. Her book *100 Plus: How the Coming Age of Longevity Will Change Everything, From Careers and Relationships to Family and Faith* was a *Washington Post* bestseller in 2011, and featured a foreword by her friend, the venture capitalist Peter Thiel. Arrison believes we don't need to passively age anymore. We can solve our bodies with the second dominant paradigm in the Valley: Claude Shannon's 1948 model of how information decays – information theory.

Most of us think of information as 'stuff you know, or if you don't, you could'. But to mathematician Claude Shannon, information was a way that you could see, predict and counteract inevitable entropy. At its most basic, it can be categorised into two realms: the mechanics of information and the contents of information. The contents is the easy part: what is being sent (say it's a love letter – this is the 'I love you' bit). The mechanics is more complicated: it's the framework that measures the probability that a message will reach the recipient uncorrupted. What might interrupt it, and how much might that affect what arrives on the other side? Perhaps you wrote your love letter in water-soluble ink, and that day it unexpectedly rained. Normally the mail carrier has a waterproof bag, but because the shower wasn't mentioned in the morning's meteorological forecast, which the mail carrier watches religiously, the mailbag got soaked. By the time it landed on your beloved's doorstep, all the ink had run and the message they received was 'I o u'. Some, but not all of this, could have been predicted.

Here's another example: a game of Telephone. Someone starts a round by whispering, 'The hat is blue', and by the time it gets back to them, through the misinterpretation of six players, the message is 'Jeremy Bearimy wearing a shoe'.

To a devotee of Shannon, information is an objective, measurable thing that can ultimately be represented as a binary digit (a 0 or a 1), aka a 'bit'. Bits are what computers read. No one before Shannon had imagined information could be part of a mathematical formula, or could be analysed by a computer.[15]

The problem with information, though, is that, as we found in our two scenarios, it doesn't stand still. The rain, whispering, anything 'noisy' can degrade a bit. This is the case with all information: bits of light, neurological signals, genes. The degradation of information is the noise on the communication channel that can't be accounted for that affects its quality over time. Entropy: the ghost in the machine.

Shannon's equation could measure the efficiency of a communication system, and predict both the probability that a particular thing might happen to a bit and how much uncertainty we can anticipate on the other side. The more complex a system, the greater the possibility of entropy, and the longer the list of things that could affect the bit, such as signal strength, channel quality or interference (noise). And where this has already been successfully applied is on something profoundly complex: a global interconnected network of networks across which information is constantly travelling.

I can genuinely feel the frisson of excitement around information theory among computer scientists. 'There's an argument that information theory is actually operating at a more fundamental level than even physics,' Elon Musk said in 2022.[16] It fosters a safe, reliable and predictable environment because it calculates the probability of unpredictable events.

The way information bits move around the web, distributed along pathways, passing through junctions, would not make any logical sense to a human mind. These junctions redirect the bit towards its ultimate end, whereupon all the bits are put back together, in the correct order, and served up. At every stage, something could happen, so there are all kinds of mathematical models that try to optimise for the fastest pathway, because the more transfer time that exists in a given bit's journey, the more likely something will interrupt it. These models, such as chaos theory and complexity theory, are outside the scope of this book, but a class of computer scientists will talk endlessly about them.

Shannon's model makes it possible to solve entropy, at least in theory. In our love letter example, one could design all mailbags to be waterproof. In our game of Telephone example, you could give all players ear trumpets. Information theory says that by breaking things into their smallest components and translating them into easily distinguishable characters like 1s and 0s, the signal that is transmitted starts out clearer, and, as it fades – or is battered by entropy – anything that diverges from the bit can be added to the system so the original signal is amplified again. And the proof that it works is that the internet can do what it does: enable huge amounts of information to be spread quickly without losing its original meaning.

What might information theory do if it was applied to another complex system, like the human? A geneticist from Harvard Medical School named David Sinclair, along with his colleagues, has taken Shannon's theory and is doing just that: using it to describe how living things age. He calls it the Information Theory of Aging.

Sinclair made a name for himself in the mid-2000s when he became the most prominent investigator of resveratrol – a natural molecule produced in the skin of grapes, blueberries, mulberries and peanuts when they're under attack from fungi and bacteria, which slows ageing in yeast. He attempted to apply it as a longevity treatment in humans, and successfully sold the idea to pharma giant GSK in 2008 for $720 million. They shuttered their research into the molecule in 2013, folding it into the rest of its R&D arm after failed attempts at early human trials. However, the original data led Sinclair to believe that we could significantly expand our lifespans to 300–400 years.*[17]

---

* He is eager to point out that he is not an immortalist; this will happen not in his lifetime, but in the far future.

In December 2023, he and his colleagues introduced the Information Theory of Aging as part of the results of a thirteen-year international research study looking into the link between genes and ageing. They found that, in yeast and mammals, the genetic information that is stored in the genome remains intact through life, but that small changes occur to the epigenome – the mechanism that oversees how genes are expressed and regulated. Think of it like building a house: the genome is the architect, and the epigenome is the contractor that chooses the materials and leads the process. The blueprint never changes, but the contractor can get tired and mess things up.

The epigenetic changes Sinclair was interested in are like the entropy in information theory: 'Unlike the stable, digital nature of genetic information,' he and his colleagues wrote in *Nature Aging,* 'epigenetic information is stored in a digital-analog format, susceptible to alterations induced by diverse environmental signals and cellular damage.'[18]

The Sinclair Lab explains this with an analogy: DNA is the information on a compact disc. Ageing is the scratches it accumulates over the years, making it skip. Resetting someone's epigenetic status is like polishing the scratches and making the CD play properly again.[19]

'He's got this idea like, your body is a hardware and you've got these software programs running,' Arrison says to me. She sees Sinclair's Information Theory of Aging as a useful way to break our bodies into information bits, anticipate entropy, and 'reprogramme' our machines.

So if the body is data and we can figure out how to fix it when it goes wrong, why aren't we already living forever? Social scientists like myself would argue that this isn't the only thing that affects our physical selves. Life – mortality – is complicated. As O'Mahoney wrote, technology immortalists aren't interested in the social determinants of death and dying – which death and our mortal bodies exist within. So if these determinants, like socioeconomic status, levels of isolation and education level (to name but a few) aren't in the equations, how can the engineers imagine all the things that are needed to correct the broken bits of our human machines?

This is where systems theory comes in – our third dominant paradigm. Systems theory is a wildly interdisciplinary approach which is used to explain how the thing we're trying to understand fits into different aspects – from atomic to biological, social to psychological, political to cosmic. That's to say, the human body is not just a sum of its biological and endocrinological and electrical and chemical systems; we also exist within an ecological system, a social system, a psychological system, a political system, a solar system, and more. Each of these layers is interdependent in a way that makes the function of all of the objects within them much greater than if you added them all up in a causal chain.

Or, think about a cake. It is a system too. Within it are subsystems – flour, salt, baking soda, butter, eggs, sugar, vanilla. These interact within another system – the oven – to create a result that's risen, melted, congealed, resisted – in a way that each of these taken on its own would not produce.

Systems theory can be used to describe the structure and behaviour of the mechanics of information theory – what happens to the message when it leaves the sender, before it gets to the recipient – because it explains the hierarchy of the different parts, as well as how they interact and how they're dependent on one another. This may seem brain-bending but, like information theory, it allows for the abstraction of everything that might be involved in a problem, and creates the possibility of producing a mathematical model to predict an outcome. Break everything into parts, simplify it, categorise it, and it can be predicted. Identify the core problem, the context in which it is embedded, and the dynamic changes that happen. Easy-peasy.

Now, admittedly, no human brain can know this kind of information for even the simplest model organism, but this is where the body-as-machine metaphor has arrived at the digital age. As we saw with GFT, computers are able to identify patterns that exist in vast datasets faster than we can, and they're getting better at doing this all the time. Data has become the organising principle for humans, rather than simply a collecting exercise.

'Consider the body as being a learning system,' wrote software architect and AI systems developer Carlos E. Perez in 2019.[20] He was suggesting a way to conceptualise life in the same way he conceptualises how large language models – like those underlying the chatbot ChatGPT – conceptualise information. We, like the machine, begin with no reference points, but we, like machines, have pure functionality: we have sight, hearing, taste, touch and smell. Our bodies continuously learn and adapt over time based on the data they collect from various systems we inhabit. Perhaps it might be psychological: developing a coping mechanism. It could be molecular: a change in our epigenome. It could be social: adjusting how we speak at work compared with when we are at home. These adaptations might be beneficial, or they could be destructive; the point is, our bodies 'learn'.

Perez recognised that if an engineer who was interested in re-coding bodies so they didn't die wanted to translate the 'ageing' bodily experience into information bits – like one would do in information theory – they would very likely miss many of the automatic things we aren't even aware of. But if they imagined the body to be a 'learning system', a computer might be able to figure it out.

What are the possibilities if a machine was left to define its own source material and do something called 'deep learning'? If it created its own connections between information? The more these bits of information are activated by feedback from systems, the stronger that connection becomes. 'I suspect that the generative models that read the epigenome are similarly adept at reconstruct- ing models despite errors in the original encoding,' wrote Perez.

But wait: this is all being done by people who suffer from Engineer's Syndrome, and an uncritical faith in their technolog- ical methods. For example, Elon Musk promises to do this with Neuralink, his brain–computer interface start-up: he has said the company is creating a chip that will record and simulate brain activity to 'give people superpowers'. 'It can also solve, I think probably, schizophrenia, if people have seizures of some kind,' Musk said. 'It could help with memory.'[21] It could. But first you have to figure out what schizophrenia and memory are, as well as how brain activity maps onto physiological, biological and psychological processes.

The irony is that the mechanistic metaphor that served us so well is now dramatically impeding further progress. Too much faith in data and engineering overlooks the value of the unknown and the unknowable. In order to defeat ageing and death, we must bend to the technical tools that are supposed to serve us – from spreadsheets to large language models. We must become more like appliances. 'This impulse, this motivation, this moral mandate to want to improve yourself means we must become like a machine,' says Dr Elke Schwarz, political theorist at Queen Mary University of London.

Yet 'we live an inconvenient life,' says Schwartz. 'We are weird. We are messy. Our bodies are mortal. We die. Why can't we be like products? Why can't we be like the things that computer scientists make that they can improve and fine-tune?' Because we aren't. So why is this the starting point for how Silicon Valley intends to 'fix' mortality?

CHAPTER 2

Start Up

I make a right turn into the first driveway I come to on Sand Hill Road, into the underground garage of the low, desert-sleek building where Shernaz Daver keeps an office as an investor, advisor, and the Chief Marketing Officer at Khosla Ventures. There are some extremely expensive cars parked here, and my budget rental sticks out like a sore thumb.

I've driven about forty-five minutes south of San Francisco, to the foothills of Stanford University, and the beating heart of the technology industry, to record an interview with Daver for the *Immortals* series. 'The billboards in Silicon Valley almost reflect the technology and the history of the valley and what kind of shift we're in overall,' Shernaz once said on a podcast, so I kept my eyes peeled at the wheel.[1] Along the freeway today, I've passed three billboards talking about living to a hundred years: one selling longevity supplements, another advertising an app for health tracking, and a third offering longevity insurance for radically extended lives.

Daver is a petite woman, glamorous in her plum suit. She manages billions of dollars for the biggest companies in tech, and I have to admit that even though I tower over her, I'm a little intimidated – and I'm not here to ask for money. 'If you're an entrepreneur, you dread coming to Sand Hill because you're going to go into big meetings with all these venture capitalists,' smiles Daver, with poise and sheen. 'But it's also a source of pride that you've actually made the cut,' she laughs. Daver studied at Stanford before the World Wide Web was a whisper, and quickly pivoted to helping the geeks that were wearing Birkenstocks at Bell Labs who were inventing the integrated circuits that were powering the internet. Over her storied career behind the scenes, she has worked with Netflix, Motorola, Walmart and Chinese search giant Baidu. She was an executive advisor at Google's investment arm, Google Ventures. She watched the sandals get replaced by hoodies when Facebook and Twitter exploded. Today, those hoodies are being replaced by lab coats. Things in the Valley are once again a'changin.'

Daver is now investing in flying cars and micro-robots to navigate the human body. These are visions of a future ten years hence, she assures me over a glass of sparkling water, as we perch in a luxe conference room.

'Venture capitalists work under the belief that they will take a risk on an idea. With the idea, with the premise that the idea is going to be hugely impactful,' she explains. 'VCs invest such that basically they get an exponential return on any investment.'

The industry is fuelled on adrenaline. Venture capitalists make lots of bets, and lose lots of money, but if they have one big hit, it can define them. Think Peter Thiel, the first outside investor into Facebook. Marc Andreessen and Ben Horowitz, the VC pair behind Airbnb. Sam Altman and any number of the successful investments at Y Combinator, the incubator he headed from 2014 to 2019. Elon Musk and Tesla.

It's difficult to overstate the cultural importance of venture capital in the Valley. Many of today's biggest funders were themselves developers who made it big during the dot-com boom of 1995–2000. When they cashed out, they reinvested their money and their mentorship. These heavyweights don't follow trends: they create them. The market follows suit. Today, they want to live forever, and they're using their influence – financial and ideological – to try to make it happen in time for them to cash in – and personally benefit. In 2022, US investments were responsible for more than 75 per cent of the $7 billion raised for longevity start-ups[2] – and though not all were in Silicon Valley, venture capital's overall investment into longevity solutions[3] was expected to be worth at least $600 billion by 2025.[4] This includes biotech and pharma companies, diagnostics solutions, supplement salespeople, and clinics.

A lot of the money comes from a small cadre of these influential fifty-something investors. Thiel, who founded PayPal before selling it to eBay and who currently has a net worth of $16.3 billion,[5] has an entire portfolio segment of biotech companies in his Founders Fund, established in 2005. Musk – the world's richest person, currently worth $342 billion[6] – was briefly PayPal's CEO, and at least two of his companies – Neuralink and SpaceX – are linked to the the immortality craze. Andreessen, net worth $1.9 billion,[7] created Mosaic, one of the first widely available web browsers, and Netscape, which he then sold to AOL.[8] He and Horowitz are currently investing in BioAge, a company which aims to target the biology of ageing with drugs that treat metabolic disorders. Altman started out by developing a social networking app called Loopt, and went on to use the profits from selling it ($43 million) to back more than 400 companies, including Retro Biosciences, which is dedicated to 'adding 10 years to healthy human lifespan' using artificial intelligence.[9] Jeff Bezos, founder of Amazon, with a net worth of $233.4 billion, joined another prominent VC, Yuri Milner, in a $3 billion investment into Altos Labs, a private healthcare company investigating 'disease reversal' and 'cellular rejuvenation'.[10, 11] Page and Brin have been investing in healthcare and longevity research for years. Brin has invested in Parkinson's research in particular; by 2024, he had spent $1.75 billion.[12] Between 2007 and 2009, Google invested $6.5 million in the over-the-counter DNA analytics start-up 23andMe, which appealed to Google not only because it was founded by Brin's then wife, Anne Wojcicki, but also because it hoped its test could identify genetic predispositions and anomalies that would help people avoid and treat disease, and extend their lives.[13]

Vinod Khosla co-founded Sun Microsystems, and is currently worth $8.2 billion.[14] He started Khosla Ventures to invest in experimental tech. 'We tend to be the first cheque into companies,' Daver explains to me. One start-up she's working with is Loyal for Dogs, which is developing drugs to extend the lives of our canine companions. Their goal is to get regulatory approval for a treatment for ageing passed through the US Food and Drug Administration's veterinary track, so they can piggyback on the precedent when they submit their application to test ageing treatments for humans. Another company, 10x Genomics, is trying to 'master biology' by mapping every cell in the body. Daver calls it 'solving' 40 trillion cells.

'We're the dreamers. We're the mavericks. We believe in the impossible,' Daver enthuses, sipping and smiling through the bubbles. 'Technologists want to build something that is going to solve a big thing. And that makes us get up in the morning.' Right now, the alarm clock for Daver and many other VCs is solutions that will mean longer, healthier lives.

A lot of people pinpoint the start of Silicon Valley's obsession with immortality to October 1990, with the launch of the Human

Genome Project (HGP), a landmark international research effort coordinated by the US National Institutes of Health and Department of Energy to sequence human DNA, generating a blueprint of how the four building blocks of life – the A, C, T and G – are arranged. Before the HGP, the biomedical establishment had used technology to sequence genomes for bacteriophages – viruses that attack bacteria – but otherwise they were peering into the unknown. The 'wet' scientists working with liquids and biological matter in labs recognised the value of computing. In return, computer scientists caught gene fever, accelerating the HGP with an injection of money, machinery and ambition. Out of this partnership came an incomprehensible amount of new tech: sequencers, analytics, computation algorithms; start-ups blossomed like mushrooms on a log in a rainforest. A whole new field was born to join biotech and bioengineering: bioinformatics, combining computer science and biology to collect, store, analyse and interpret data. In 2003, two years ahead of schedule, the human genome was sequenced, 'in part', according to the US government's National Human Genome Research Institute website, 'due to a deliberate focus on technology development'.[15] It shook the foundations of society (raising concerns about genetic discrimination and privacy), philosophy (raising questions about the role of genes in shaping individuals and societies, and the nature of identity), politics (influencing funding priorities and legal frameworks) and medicine (paving the way for new ways to diagnose and treat disease). And it opened a new vertical in the Valley.

'I went over one night to a friend's house – a very talented hacker,' Sonia Arrison says as we sit in her sun-drenched conference room on Sand Hill Road, a few properties down from Khosla Ventures. 'He had *Intro to Biology* books all over his living-room floor,' the investor tells me. 'I looked at him, and I'm like, what is this about? Are you planning on changing careers? And he's like, Sonia, no. Today I'm coding ones and zeros. Tomorrow I'm gonna be coding the A, C, T and G of human DNA.'

As well as funding a longevity drug research start-up,[16] Arrison has invested in an AI-powered personalised health diagnostics app,[17] and a nutrition company developing a product that 'focuses on activating cellular reprogramming and rejuvenation to support functionality and health in multiple organs and systems.'[18]

'Once one of these rejuvenation technologies hits the market, it's going to spread faster than ChatGPT,' she tells me, her blonde hair bobbing enthusiastically as she leans closer. 'People will eat this up like crazy. No marketing team will be needed, I'm telling you. It's gonna be – to use the Silicon Valley term – viral.

'There was a time not long ago when the idea of reversing ageing sounded like quackery,' she continues. 'Maybe it even does today to some people. But scientists have proven that ageing is malleable. It can truly be reversed.'

But have they, really? It depends on who you ask.

In the 1980s, Aubrey de Grey was working towards a computer science undergraduate degree at the University of Cambridge. After graduation, he eased into a job at Sinclair Research creating early AI, but a serendipitous encounter at a party with a fruit fly researcher – and his future wife – took him on a very different path.

Adelaide Carpenter's boss at Cambridge was looking for someone who knew about both computers and biology to oversee a database about the insects. De Grey took the job. For more than a decade, he was the software developer for the Genetics Department, building and maintaining the FlyBase dataset – *the* primary repository of genetic and molecular information about the model organism, the fruit fly (or *Drosophila*).

De Grey spent his free time deep diving into ageing, reading all the literature and attending all the conferences he could. He published a book on his research in 1999 called *The Mitochondrial Free Radical Theory of Aging* which won him a PhD by publication in biology and a place in the pantheon of immortalists.

There is no way to describe de Grey without mentioning his beard. It's long and wizard-like and, increasingly, a Gandalf shade of grey. It's also difficult not to mention that he is certain that the first humans to live to 1,000 years have already been born. In 2011, he said they were probably already sixty years old.

De Grey is a controversial figure in the research science community. This is in part because of his views – he believes in the power of regeneration – but also because in 2021, he was accused by two women of sexual misconduct and, after being found to interfere with the investigation, three of the four allegations were upheld by an independent report.[19]

Yet he remains in the public eye when it comes to longevity research. In the lab, his research is laser-focused on creating technological and biological solutions for age-related physical and cognitive decline. He's reportedly been granted millions of dollars in investment from Thiel and other Silicon Valley billionaires and his reputation draws idealistic and intelligent researchers from around the world to go to work for his research foundations in Silicon Valley and in Cambridge. His message is thrilling: radically extended lifespans will be with us in less than a decade, and through his own lab near the Santa Cruz mountains, and the various foundations he has established, he is pouring money and energy into slowing and reversing cellular ageing and breakdown. His plan is to kick-start an anti-ageing industry by doing proof-of-concept research on things that are too early to commercialise, except by 'the most visionary investors'.

In 2009, de Grey suggested that 'the high-profile academics

who occupy the pinnacle of opinion-formation' are 'not terribly interested in doing anything about aging'[20] This is arguably a point of difference with those 'high-profile' researchers studying ageing, like Nobel Prize-winning structural biologist Venki Ramakrishnan: 'De Grey has learned enough about biology to pinpoint many of the things that go wrong as we age,' wrote Ramakrishnan in his 2024 book *Why We Die: The New Science of Ageing and the Quest for Immortality*, but 'he is wildly optimistic about the feasibility of addressing them.'

De Grey likes to think of himself as an outsider, free from institutional politics, so he's not afraid to stir the pot, he tells me when we speak on Zoom in March 2025. As the rain pounds the window in New York and the sun shines through de Grey's window in the foothills of the Santa Cruz mountains, he tells me he thinks his approach to age research is responsible for popularising the idea that people don't have to age: 'I have kind of dragged the field kicking and screaming, well . . . the whole world kicking and screaming into a reasonable, a somewhat better, you know, level of respect for the idea that ageing is an undesirable, and potentially medically amenable, phenomenon.'

Ever humble, he tells me he considers longevity researchers to be his peers – 'you can hardly put a piece of paper between me and the people you would call the geroscientists,' he says. But 'there is a significant difference between me and, let's call it, the centre of gravity of the geroscientists, with regard to optimism. In other words, with regard to how soon we think a certain amount of progress is likely to be made.'

De Grey has organised his career around getting things moving. Back in 2003, frustrated with the slow rate of translating the findings from the lab to treatments in the field, he attended the American Aging Association's annual event representing the

Methuselah Foundation, the non-profit he co-founded that aims to make '90 the new 50 by 2030'.[21] He particularly caught the attention of gerontologists and journalists when he announced the non-profit's first big competition, the Methuselah Mouse Prize. They would award £20,000 to two teams: one which broke the world record for the oldest-ever lab mouse; and one which developed the most successful late-onset rejuvenation strategy. The stunt got global headlines, and the winners – a research team from Southern Illinois University whose mouse, GHR-KO 11C, lived for 1,819 days, and a researcher from University of California, Riverside, who extended the lifespan of middle-aged mice by 15 per cent – split the purse. Two years after announcing the Mouse Prize, he spoke at the first TEDGlobal conference and his talk, 'A Roadmap to End Aging', has since been watched 4.8 million times.[22]

De Grey has focused on building out his research framework, Strategies for Engineered Negligible Senescence (SENS), known at the Methuselah Foundation as 'getting the crud out'. The idea is to reverse or prevent the damage at the molecular or cellular level that happens as a result of ageing. Environmental factors might play a part, for example, like air pollution and UV light.

This has been an effective message to the can-do VCs in Silicon Valley. In 2006, Thiel gave de Grey's team $3.5 million to reverse the signs of and debilities caused by ageing. A year later, de Grey published his second book, *Ending Aging: The Rejuvenation Breakthroughs That Could Reverse Human Aging in Our Lifetime*, and was interviewed on the US current affairs programme *60 Minutes* and by the *New York Times*. In 2009, he span the SENS research track of Methuselah out into its own foundation, where it has funded research projects at academic institutes. For a long time, de Grey was Silicon Valley's chief – and only – gerontologist. But in 2013, Google announced a brand-new initiative: their own spin-off company that would undo ageing.

∞

GFT had demonstrated not only the size of Google's dataset, but also the technological might it had to make connections. The company's databases had swelled to even more text, images, video and sound, as well as details about our social world. For an engineer, it seemed obvious: the data about diseases of ageing was simply knowledge that hadn't yet been discovered.

The new company, Calico, would be devoted to 'moonshot thinking around healthcare and biotechnology', wrote Larry Page.[23] Sceptics immediately denounced the hubris of the venture, believing it was a tech god's attempt to discover the fountain of youth. It likely didn't help that Page admitted in the sole interview he gave about Calico that he wasn't an expert in diseases of ageing, but did have knowledge of the tech involved in medical research from 'just being in Silicon Valley'.

'Are people really focused on the right things?' Page said to *Time* magazine. 'If you solve cancer, you'd add about three years to people's average life expectancy. We think of solving cancer as this huge thing that'll totally change the world. But when you really take a step back and look at it, yeah, there are many, many tragic cases of cancer, and it's very, very sad, but in the aggregate, it's not as big an advance as you might think.'[24]

De Grey was delighted when Calico launched. He announced: 'With Google's decision to direct its astronomical resources to a concerted assault on aging, that battle may have been transcended: once financial limitations are removed, curmudgeons no longer matter.'[25]

And indeed, their resources were astronomical: their initial
investment was \$1.5 billion.[26] Google's market cap at that time
was just shy of \$300 billion – so this exclusive funding repre-
sented 0.5 per cent of their value. Calico hired Daphne Koller,
computer scientist at Stanford and co-founder of education
platform Coursera, to be their chief of computing. Biology was
being transformed 'into a data science,' she said in a statement
announcing her hire. 'But the potential of the massive data sets
that are and that could be produced will only be fully tapped via
the development of powerful computational tools.'[27]

The company stacked their team with molecular biology
and genetics superstars, able to attract top researchers with
sports-star salaries and a mandate to tackle 'the most fundamental unsolved problem in biology.'[28] The small first cadre of
staff included famed biochemist and Apple board member Art
Levinson, who was poached from goliath research and development organisation Genentech to be the research lab's CEO,
and longevity wunderkind Cynthia Kenyon, who had made her
name in the ageing field at the University of California, San
Francisco, appointed as Vice President of Aging Research. De
Grey was expecting big things.

But by 2019, there was no radical treatment for ageing that
came out of the six years of research. In fact, most of the news
coming from the Calico camp was about partnerships with
world-class laboratories, which were focused on basic science –
the research necessary to understand the problem, rather than
the research required to translate the science into a treatment.
Calico was not a Silicon Valley investment fuelled by adrenaline,
focused on releasing a product. No one was more surprised –
and disappointed – than de Grey. He began to grumble about
the initiative, calling it a 'clustershambles.'[29] By the time I spoke
with him in 2025, he had grown even more disappointed by

Calico: 'Calico is an absolutely monumental catastrophe.' Why is that? 'Because they didn't listen to me, basically ...'

As he rants, I realise that though I sense de Grey is annoyed he wasn't hired, he's actually more disappointed that Calico wasn't the success story he and his brand of immortality were hoping for. 'They needed to hire someone to bridge the gap, you know, to get from proof of concept to concept. And that means a chief technology officer, like, for example, me ... So, it will be a complete miracle if anything of any significance ever comes out of Calico.'

Today, de Grey is only one of a number of gerontologists available to Silicon Valley, and the landscape has changed. There are now many more biotech companies around than when de Grey launched the Mouse Prize, and the biggest names with the biggest investment, like Altos, or the Buck Institute for Research on Aging, are slow to take their science to human trials.

The exception, de Grey tells me, is Retro Biosciences, founded in 2023, initially funded by OpenAI's Sam Altman with a staggering investment of $180 million – the largest single investment in a longevity start-up to date.[30]

Joe Betts-LaCroix is the co-CEO of Retro, with Altman. He introduces himself to me as a computer scientist by trade who wants to get in on the ground floor so he can make a difference. 'What are the things that cause people stress, strain, fear, insecurity today that I actually can have an impact on?' he asks me on the upstairs deck of a tech start-up event space near San Francisco's Mission District. 'One of the biggest ones that also has a technological route is people getting sick and dying.'

He tells me that his role in longevity research is to ultimately reduce people's insecurity. 'Like, people are walking around all the time thinking, damn, I'm gonna die. I better hurry up and save enough money so I'm not too uncomfortable during that horrible dying process between when I retire and when I start falling apart.' And so Retro is working on building cellular therapies to 'meaningfully reverse age-related diseases'[31] – exactly what de Grey has been preaching for years. 'What kind of good can come from lifting some of that weight off of people?' Betts-LaCroix posits. Downstairs, people are milling around and eating canapés at a longevity networking chinwag.

He's motivated and ambitious, and is positive his company can do it; he tells me the company aims to have a drug in trials in 2025. That March, they announced the first round would be starting in Australia. This is more de Grey's speed. This is the kind of immortalist attitude that will get things done by leapfrogging the establishment and cashing in on disrupting life.

CHAPTER 3

Vitality, Inc.

For people lose all appearance of mortality by living in the midst of ✗
immortal blessing.

Epicurus

Bryan Johnson was born in 1977, but according to his doctors his body is much younger.

That doesn't mean he looks like a teenager. It's difficult not to stare, to try to figure out what's not quite right. His unnervingly smooth skin is almost translucent; there's no fat anywhere on his body. His eyes – clear, blue, piercing – are the only things that move when he's listening.

Johnson claims to have found a way to make time stand still. ✗ Based on his physical metrics alone, you could be forgiven for thinking he's cracked it. His routine measurements of health, strength, physical resilience, and endurance are encroaching on medical-miracle status. He's a pincushion for frequent blood donations, a regular at the MRI lab, and an early adopter of health-tracking gadgets: a WHOOP band on his wrist, an Oura smart ring, a Pulsetto de-stress device – a vagus nerve stimulator – that he uses to regulate his nervous system, other brain stimulators he wears around his neck and in his ear to help him remain 'chill', and a breath-tester gizmo to track lung health.

All of the data these tools provide is automatically uploaded to an iPad and this data determines how he lives his life. I call him a lab rat. He prefers guinea pig. To his friends and family, he's 'Bryan' or 'Dad'. To the world at large, he's Bryan Johnson: anti-ageing explorer, radical life extensionist, biohacker, and professional rejuvenation athlete.

'I propose this situation where my body is going to be integrated with science and technology, and it's going to run itself,' Johnson explains to me via Zoom from his beach house in Venice, California.

'If you think about it from an investor perspective, most of the time investors say, if I invest in this company, I'm going to get a three x return on my investment or a five x or ten x on capital. People are connecting the dots and saying that if one were to dedicate money and time to this area, you can expect meaningful returns. If one invests in their lifespan right now, you may get a ten x return on your life. You may get an extra ten years or a hundred years, actually, or a thousand.' That's a lot of extra life.

The way Johnson invests is through dedication to what he calls his 'Protocol': he wakes up naturally around 4.30 a.m., takes three supplements, does 3–5 minutes of natural sunlight therapy (mimicking standing outside in the California sunshine), checks his temperature, does in-ear electrical impulse therapy for more vagus nerve stimulation, swallows fifty-four supplements, drinks his 'green giant' concoction, wears a laser diode hair cap, works out for an hour, and then eats 'a few pounds' of vegetables. Then he does electric stimulation on his abs, 12 minutes of near- and red-light therapy, hearing regeneration therapy, eats a calorically rich, nutrient-dense serving of 'Nutty Pudding', and then takes thirty-four more supplements. Like other influencer biohackers, his 5–9s are as important as his 9–5s.

Throughout the rest of the day he does a series of other

rituals, including bowel imaging to check his gut health and penis shock therapy (to 'grow new blood vessels and improve blood flow'),[1] all while liaising with his medical team to assess any new treatments they're thinking of bundling into his wellness routine, building his brand on social media, strategising his lifestyle supplement company Blueprint (which also sells test kits, and discounted access to MRI scanning for up to $5,700),[2] and running the Rejuvenation Olympics – a series of leaderboards (requiring a diagnostic subscription that costs $500 per test or a discounted rate of $1,000 per year) for people to post their data and find out just how longevity-fit they are.

Johnson's lifestyle is powered by the 'fundamentals' of bio-hacking: goal-oriented, natural and technological approaches to a hyper-optimised lifestyle and mind–body connection. Around this he has built his own subscription service for his Blueprint-branded lifestyle hacks, a product line of Blueprint-branded supplements and foods, and a Blueprint media presence, with 1.7 million subscribers and more than 250 million views on YouTube, over 2 million followers on Instagram, and over half a million on X.

Data is Johnson's fuel. But he isn't alone. There is a movement of data-watchers who fundamentally believe they are their best selves when the numbers say so. Johnson's competition in the field of biohacking is the 'Father of Biohacking', Dave Asprey, who started his own wellness empire in 2012 with his Bullet-proof range of products, supplements and foods, alongside the Bulletproof self-help enterprise. Asprey defines biohacking as 'the art and science of becoming superhuman'. The biohacker's job is to 'change the environment outside of you and inside of you so you have full control of your biology, to allow you to upgrade your body, mind, and your life'.[3] Tim Ferriss is another superstar health guru with videos and products. Ferriss offers his followers 'life hacks' in bestsellers like *The 4-Hour Work Week*, *The 4-Hour Body* and *The 4-Hour Chef*. Life-optimisation 'commitments' and other investments – including Facebook, Uber and more than fifty other companies – have netted him an estimated $100 million.

In 2024, the biohacking industry was worth around $24 billion – projected to grow to $28 billion in 2025 and $133 billion by 2034.[4] Johnson, Asprey and Ferriss share a DIY approach to self-care, which means they are endlessly marketing tools to monitor every biological function. The aim is better, longer lives, and salvation through optimisation.

The real money-making opportunities lie in the ever-evolving evidence these men claim is fact. There is never consensus on what actually works to be ever more healthy, and biohackers are forever debating what the most effective routines and interventions are for total wellness. Some say more sex, others say none. Some say drink red wine, others say don't. Some say sun exposure is good, some say it's bad. Tribes congregate around the lifestyle that resonates with them, and around the persuasive gurus who sell their user guides to 'optimal human resilience', as well as their extensive product range of coffee, cookbooks, webinars, showerheads, biomedical testing, online courses, monitoring devices and food – so followers can live like 'Titans'.[5]

Before he caught the lifestyle bug, Johnson was a serial entrepreneur who'd dabbled in cell phone sales and real estate. In 2007, when he was thirty, he landed on the credit card processing technology that made him a centimillionaire: Braintree. His e-commerce company helped businesses receive money digitally, thus removing the intermediary from transactions: it was a person-to-person solution. Five years later, in 2012, Braintree was handling $1 billion in e-commerce on cell phones – a whopping 10 per cent of all mobile transactions. The company purchased mobile payments leader Venmo for $26.2 million, combining its customer-facing technology with Braintree's business-side solution. PayPal bought it for $800 million one year later.

Johnson's story of transformational wealth fits in nicely with others from the disruptor class of Silicon Valley. A boy from humble beginnings in Utah, he grew up in the Mormon church with his mother, stepfather and four siblings in a middle- to low-income household. He had only intermittent contact with his biological father from a young age, after his parents divorced and his dad left the church. For the first thirty years, everything in Johnson's life centred around his faith.

Within a year of selling Braintree/Venmo to PayPal, he got divorced, leaving his wife and three young kids, and his religion.[6] Fearing he'd lose touch with his own children as happened when his own dad, Richard, left the faith when Johnson was a child, he tried to balance being a divorcé dad with the gruelling expectations of being an entrepreneur. He started another company, and descended into depression.

In May 2018, Johnson wrote a 4,800-word manifesto for the future of humanity. In it, he proposed a plan for our collective survival: first, admit we're flawed. Second, evolve. To do this, we have to radically improve ourselves through any means necessary – supplements, meditation, exercise, self-help. We also have to digitally mine ourselves – turn everything into 1s and 0s – so that we remain relevant in the future and can continue transforming: 'If we want to ensure we all get as many tomorrows as we want, it's time to get to work,' he wrote.[7]

On 6 August 2020, Johnson 'fired' his alter-ego, 'Evening Bryan'. Evening Bryan was chronically tired and frequently miserable. Evening Bryan retreated into a tray of brownies after a hard day at the office. Evening Bryan felt bogged down by expectations.

The transformation wasn't easy. He once described it like peeling back twenty layers of an onion: first the brownies went, then the pizza parties and other ultra-processed foods. Next, he addressed his poor sleep hygiene, and then he tackled his psychological self. He implemented a strict formula of meditation and exercise. Soon, he realised he had to let 'Confirmation Bias Bryan' go. But by the time he got to the onion's core and his all-Bryan meetings were just him on his own, he realised that these tweaks weren't enough. To radically extend his life, he needed to do something even more transformational. He needed to delegate his mind to data.

Today, he is no longer a Mormon, and he lives on his own near the ocean in Venice Beach, Los Angeles, a part of town where everyone seems to float slightly above the earth rather than walk on it. 'I love you,' says the woman who hands me my bill in a restaurant around the corner from Johnson's house, after serving me a plate of something raw listed on the menu as 'I Am Beautiful'. Everyone around here seems to be searching for something, and to have come to the land of sunshine to find it. Around the corner at the original Gold's Gym, a global destination for bodybuilders since it was opened in 1965, people are doing reps flanked by crews filming them for their social media channels. Influencers are taking selfies. I see lots of wearable tech, and lots of people logging their lives.

But it occurs to me that none of this obsessive recording is new. Long before digital tech, we also logged our lives. We kept diaries. We wrote memoirs. Stretching back to the 1700s, we used journals with prompts – like today's gratitude and bullet journals – and agendas to take our 'moral temperature' and calibrate our 'moral compass'. Logging life has been used as a way to better understand ourselves: we create records for us to return to, and reflect on. For many, being able to keep track of the minutiae has been a way to keep ourselves virtuous, and to know we're alive.

During the first two decades of the twenty-first century, this practice evolved into the Quantified Self movement, which took *usa* the same motivations – self-knowledge through self-tracking – and quantified it – 'self-knowledge through numbers'[8] – using digital note-taking software, blogs and data visualisation tools.[9] Nicholas Felton's beautifully designed, infographic-rich lifelogs from this time, called the Feltron Reports, are a notable, super-powered example of this in practice: five years' worth of them are in the permanent collection of New York's Museum of Modern Art.[10] I interviewed Felton for an episode of BBC Radio 4's *The Digital Human* in 2014 about our drive to turn ourselves into numbers.[11]

'I think that my brain is a combination of a designer and a scientist, and so I like to understand the world in quantified ways,' he explained. Between 2005 and 2014, Felton documented everything he did – not just what he ate and how far he ran, but how often he slept alone, how many times he went to the movies, how many flights he took, which books he read, how many miles he drove, who he spent time with, and much more. His work translated directly into the digital space: he had a hand in shaping the Facebook Timeline design,[12] and it was thanks to his reports that he got the job.[13]

For social researchers, reality TV watchers, students of human behaviour, designers, and anyone else interested in people-watching, the Feltron Reports were a compelling study in personal data – after reading about how many photos he took with his iPhone and how many airports he'd visited in the previous year, plus all the hundreds of other categories of itemised frequencies, I felt like I had a real insight into who he was as a person, even though I'd never met him before our conversation. For a while, he personified the practice of collecting personal data, and seeing what new information rose to the surface; he released an app called Reporter, which gave users the tools to record the subtleties of their lives and gain insight into unexpectedly important things through beautiful graphs that visualised the data.

The value that came from quantifying the self was seeing himself reflected back in the numbers. What Felton did was above and beyond, mostly because he tried to capture everything. This really wasn't possible for everyone. But in 2009, a device arrived on the scene that ambiently captured data about its user in one category – health data – making this self-tracking possible for anyone who wanted the insight, but didn't want to do much to get it.

Fitbit is a wristband device that in its earliest model solely counted steps and how many calories a wearer burned. It was a runaway hit and dominated the market for a decade, increasing in functionality with each model and creating a whole new industry of gadgets with embedded electrocardiograms and biosensors that read physical signals and extract clinically relevant data. For many people who bought the devices, this medical function fell outside any need to know, and was more often a curiosity. Over time, fitness trackers have come to be recognised by some as a necessity; in the US, some health insurers distribute them to their clients, or give discounts to people who share them.[14]

These affordable technologies have been spun out in the same way the social networking pioneers promised to change how we connect: heavy on futuristic scenarios, messianic pronouncements, and idealised stories about transformational outcomes. And they seemed to get results. The big trend in app design when the Fitbit was first released was 'gamification', designing software with game-like elements (rewards, reminders, tasks and missions) even in non-game contexts. Engineers programmed systems to save the princess, eat all the dots, take 10,000 steps a day, run a 10k, and the user didn't realise the hard work that was going into the outcome. Over the last two decades, hundreds of start-ups have cashed in on the personalised wellness craze. Their price points aren't aimed at the extremely wealthy, either; self-tracking is now within reach for many. The global revenue of this industry was expected to hit $46 billion in 2025.[15]

As the Quantified Self movement was creeping into positive health outcomes, designers realised that data could be used to influence and persuade, on a very large scale. The UK's Behavioural Science Unit, informally known as the 'Nudge Unit', began to use data science to gather insights and direct the public towards choices that encouraged behaviour in line with government policy – paying taxes on time, getting COVID-19 vaccines, turning up in court, increasing organ donations.[16] What became the most important feature in implementing new interventions or releasing new products across sectors was what kind of data could be gathered to provide what kinds of insights.

An unanticipated consequence was that tracking data didn't just change behaviour, but what the outcome meant as well. Take health data, for example. In the past, we didn't need to know our ECG or our EDA (electrodermal activity), but now that these have been quantified and attached to our wrists, there's a number that describes 'good' and one that describes 'bad'. But do either of those numbers describe whether we are or are not 'well'?

Today we are swimming in conflicting and ambiguous messages about wellness, which translates into a global wellness industry valued at $6.3 trillion. Not counting creams, pills and injections, the 'basic' wellness protocol – healthy eating, nutrition, weight loss and physical activity – is worth just over a third of this, or $2.2 trillion.[17] Tech health solutions mean it's never been easier to track just how healthy you are – and how healthy you are not. But there's no agreed definition of what wellness actually is. The Global Wellness Institute defines it as 'the active pursuit of activities, choices and lifestyles that lead to a state of holistic health'. The global pharmaceutical company Pfizer says, 'the act of practicing healthy habits on a daily basis to attain better physical and mental health outcomes, so that instead of just surviving, you're thriving'. But because 'holistic health' and 'thriving' are subjective concepts, being well for those who are actively choosing it is always just out of reach, and achieving it has become borderline obsessive.

'We no longer avoid sinning for fear we'll be shut out of heaven; instead, we avoid unhealthy behaviour for fear it will make us sick,' wrote feminist author Ione Gamble in the *Guardian* in 2022.[18] In January 2025, there were more than 25 million downloads of the leading fitness and workout tracker apps worldwide, up from 23 million downloads a year before.[19] Fitbit was the third most popular app, after MyFitnessPal and Strava, generating $7 million that month alone.[20] As many strive for the ideal dictated by the #fitspo hashtag,* or the numbers the biohackers post about their nightly erections, they follow their moral signposting, and wellness becomes how they live their lives – and how the fitness gurus pay their mortgages. Yet there is evidence that tracking is creating a shift in what we consider 'healthy'. According to Michał Wieczorek and his colleagues

---

* 'Fitspo' is short for 'fitspiration', a buzzword for online content meant to motivate people towards being physically fit through rigorous exercise, diet and lifestyle. Many wellness influencers and others have consolidated around this social media hashtag, which has been criticised for being too focused on measures of physical attractiveness.

writing in the journal of *Public Health Ethics*, health tracking has become a performance. To be healthy is to be virtuous, they write – and the numbers are there to prove it.[21]

Wellness and health have for a long time been in an uncomfortable clinch with virtue. According to scholar the Rev. Dr Melanie L. Dobson, this stretches as far back as Thomas Aquinas in the thirteenth century: 'health constitutes not just a status for Aquinas,' she wrote, 'but a moral activity in which every person participates ... When we care for ourselves through habits of health, we become better people; such improvements are a part of Thomas's design for the virtuous life.' Dobson goes further, suggesting that Aristotle's description of the good life is intertwined with healthcare.[22]

And so it follows that there have also always been fads that lean into this sense of wellness as a moral objective. At the end of the nineteenth century, the then priests of the 'hygienic religions' monetised and moralised that being 'well' was the 'good' response to a chaotic world of 'bad' influences (dare I say a 'sick' society) and oversaw wellness fads similar to those of today's biohacking gurus. John Harvey Kellogg was the biggest #fitspo influencer of his time, heading up the Battle Creek Sanitarium in Michigan – originally a Seventh-day Adventist institution founded on the principles of hygiene revealed not by science but by God himself. Between 1874 and 1953, Kellogg published hundreds of health pamphlets – that era's blog posts or YouTube shorts – extolling vegetarianism and decrying fashionable dress.[23] He cashed in on the craze with his own breakfast cereals, nut butters and nut-based meat substitutes to boot.

By following him, anyone could achieve a secular state of grace. It just took self-discipline and self-control to eat healthy foods, do specific exercises, perform the recommended routines, and spend weeks at a time in his sanatorium. Home health hackers could buy Kellogg's meat substitutes, Nuttose and Protose, and other staples of the 'Battle Creek Diet System'.[24] Longer lives could (theoretically) be available to all – though the super-dedicated and the super-wealthy had a better crack at the prize.

Today, full-body scans are the new status symbol, wrote the *New York Times* in 2023.[25] It's only after getting the baseline data, the biohackers say, that you can predict your future. Both eras have quick fixes, such as yogurt enemas, sexual abstinence and vegetarianism. It doesn't matter whether we are indulging in Nuttose or Nutty Pudding, if we are following a wellness regime, we are signing up to a moral code.

Some social theorists believe that there's a degree of psychological projection going on with fitness disciples and biohackers, that they are responding to an unhealthy culture. By embracing the dogma of health, they are doing a public good: sacrificing their bodies in the pursuit of human perfection for all. 'This is why I'm really doing this,' Johnson tells me as the sun streams through his window. 'I'm currently trying to retell the origin story of the human race.'

It's forgivable to think that this is absurd, even in Venice Beach.

'If we're in the twenty-fifth century, and we're observing what people in the twenty-first century figured out that allowed for the continuation of the human race, we may say they had three problems,' he explains. 'One is they were on the doorstep of building superintelligence. Two is their biosphere was in question. And three is they were just dangerously at each other's throats, you know. Fingers on the nukes, bioweapons. Like they just had this violence towards each other. They were always just one step away from catastrophe. But they figured out cooperation at levels that no generation of humans ever figured out.

'We no longer charge battlefields with swords and bayonets,' he continues. 'We have the rule of law. We have moved to these non-violent means. We can resolve conflict through these increasingly sophisticated means.

'If we want to take this to the next level, I wanted to be the individual representation of that.'

Biohacker Bryan is trying to solve the problems of the world by living according to his 'Protocol', run by his Blueprint algorithm. It is this that he believes will get his 35 trillion cells to cooperate with one another.

'When my body, when my mind is like, let's go eat pizza or let's go drink . . . that's non-cooperation. I relabel those things to violence,' Johnson says. Violence that accelerates ageing, the onset of disease, and ultimately death.

'You're literally lessening your lifespan by doing these things,' he insists. 'I may come up with a pretty story of like, this is the way I live life and YOLO, and whatever cute narrative I want to come up with. You can't get away from this idea that it's legitimately violence against self.'

Johnson's objective is 'zero violence'. If he can do that, he says, his Blueprint algorithm is equally applicable to planet Earth. 'What did the people of the twenty-first century accomplish?' he asks again. 'They figured out how to computationally cooperate at a scale never before seen.'

'Unwellness' may be an individual problem with an individual cause and an individual solution, but by publicly serving as a model for humanity and computing cooperation, Johnson hopes he can lead us all to what he sees as a global good: the abolition of death.

He's not the only biohacker who's shifted from wellness in general to the more elusive eradication of ageing; in the last few years, the most vocal have brought that crusade front and centre. Ageing is stereotyped as negative: in Dave Asprey's original concept for Bulletproof coffee and the Bulletproof diet – which he launched in 2013 as 'Bulletproof 360' – ageing was one of eight 'systems' that should be hacked. But by 2019, when he released his book *Super Human: The Bulletproof Plan to Age Backward and Maybe Even Live Forever*, battling ageing was the brand's key focus. Taken from the perspective of the can-do, solution-oriented world of data, it is the solemn duty of a virtuous person to counter and slow the processes of ageing, preferably by using anti-ageing medicine and cosmetic surgery.

Wellness has morphed into a way to treat deviance from 'normal', where illness, being saggy, being old is 'wrong'.

In 2017, Sally Adee was a science and technology features and news editor at *New Scientist*. She'd reported on the anti-ageing field in tandem with the latest about stem cell therapies, human growth hormone, caloric restriction, a diabetes drug called metformin. All promising as lifespan extenders, but the proof that any of them actually worked was still pending. As a reporter, her job was to stay steady and keep to the facts, but that year, things were hard. She was in her late thirties trying to balance high-pressured deadlines and the needs of her infant twins.

'I hadn't slept in two years ... I was convinced I was at death's door,' she tells me over the phone from her office in London.

Sally was tracking a lot of the research on sleep. Matthew Walker, the Director of the Center for Human Sleep Science at the University of California, Berkeley, had just published the results of a four-year review, showing that a shortfall over the long term had connections to Alzheimer's and cardiovascular disease, among other things. It ruins the body and, as anyone who's stayed up late writing a term paper or caring for small children knows, it ruins the mind.

'I probably wasn't thinking as logically as I might have otherwise been,' admits Adee. 'But I was so tired and I felt like I had aged thirty years in the past two.'

So, in pursuit of a story and some personal insight, she got a panel of tests. She went to Cambridge to assess her cognitive functioning, to Brighton to measure the age of her skin. She underwent eye tests, blood tests and a VO_2 max mask test – a measurement of fitness that tells you how much oxygen you have on hand during exercise. She was poked, prodded and observed using the most up-to-date scientific equipment. If anything was truly threatening Sally's health and potentially curbing her longevity, this should pick it up.

But as the results arrived, it seemed that everything in her baby-tired body was functioning exactly as it should. She was 'well'. Yet she felt that the opposite was true.

Over the last century, thanks to social and public health innovations, such as the eradication of smallpox through vaccines, and a more organised medical system, health outcomes have improved dramatically. Globally, life expectancy more than doubled between 1913 and 2023, from 34.1 years to 73.2.[26] In the US in 1900, there were around 120,000 people aged eighty-five and over, representing about 0.16 per cent of the total population;[27] by 2022, that figure was 6.5 million – nearly 2 per cent of the population.[28] And by 2050, predicted the advocacy group Institute on Aging, Americans aged eighty-five and over will make up 5 per cent of the population: a whopping 19 million people.[29] There were an estimated 16,140 centenarians in the UK in 2023 – more than double the number in 2003.[30] But what does all of this extra life feel like?

'The question to ask is not what can't be measured, but what is not measured,' cautions author Amelia Abreu when I interviewed her in 2014. She was concerned that tests and trackers aren't for everyone, yet a large proportion of everyone is using them for their somatic sense of their own wellbeing. 'I think about the use of data-collection devices for tracking physical activity. What about emotional activity?' she says. 'Like, if I have a really stressful week, that doesn't show up on an activity tracker.' What's missing from the data is the 'why'.

There are other contextual things data doesn't capture either. For years I slavishly fed my fitness tracker with my exercise data, receiving badges and virtual high-fives when I got stronger and exercised more. I was levelling up in the game of fitness. The win condition was that the numbers continued to go up.

And then, happily, I got pregnant. For a while, I was able to continue as before. But then I wasn't. As my activity times slowed and my sessions became less frequent, I got pings saying I was neglecting myself and being unhealthy and, like Sally Adee, I was gravitating away from the measured state of wellness. For the record, my body was busier than ever.

'The philosopher Sandra Harding has this really great quote where she says that when you limit the definitions of what needs scientific explanation to bourgeois white males, you often get a partial and sometimes perverse understanding of things', Abreu says when I share my experience. 'I have a friend who's a competitive triathlete. The first thing she always admits is that her ability to train is limited by her access to childcare. That shapes the data as much as the activity itself.'

Admittedly, this is a bigger social issue: despite representing approximately half of the population, women are 'a deviation from standard humanity', writes Caroline Criado Perez in her 2019 book *Invisible Women: Exposing Data Bias in a World Designed for Men*. Health practices that promote the qualitative – or subjective – lived experience, and particularly the nuances of female perceptions, feelings and interpretations, have driven wellness movements online, where women have the freedom to express themselves, remain in control of their own bodies, and are empowered by being taken seriously. But this also opens the door to misinformation, spreading like folklore through these communities, which, as I reported in 2022, can increase disenchantment with public health bodies and lead to problematic conspiracy theories.[31]

Medical interventions are not optimised for women, and women are woefully overlooked in age research.[32] Health problems that mainly affect women are generally not studied in an integrated way; few medical schools offer women's health courses or include gender-specific information in the curriculum. Sex and gender affects diagnoses, treatments and prognoses; for example, women are twice as likely to experience adverse reactions to medicines. Women have also been underrepresented in clinical trials, which leads to under-development of medicines for female physiology. If this information isn't present in the research going into the development of the data-trackers, they won't be optimised for women, either. Both Sally's and my experiences of feeling ignored by the data are examples of how the externalised, numbers-driven biohacking community tends to be coded 'male'.

Where this becomes a problem is in interpretation. A friend had a baby and was deeply overwhelmed. She was recommended a baby activity tracker by her doctor to follow her newborn's behaviours as a way of asserting some control over the chaos. She religiously entered sleep times, diaper changes and their contents, and meal times, lengths and quantities. When I saw her after she'd been using the tracker for six months, she was more overwhelmed than ever. She had pages and pages of graphs that showed trends, but still was struggling. And worse, she didn't feel bonded to her baby.

I suggested stopping the trackers and stepping away from those fluctuating graphs. When I saw her a month later, she told me that letting her apps go was simultaneously the most difficult and the best thing she could have done. 'I started actually listening to the different cries,' she said. It was as if the numbers had been louder than her child, and didn't let her hear what her daughter was saying.

I don't think that the biohackers are entirely motivated to slow down ageing by a fear of death. I think they are masking something else with their obsession with optimisation. There will come a time when Johnson's, Asprey's and Ferriss's numbers will go down. How, then, will they redefine themselves?

By tracking our vitals with rings, smartwatches and epigenetic clocks, and believing that what comes out represents an objective measure of whether we are well or not, we've forgotten what well feels like. We've lost touch with the ebbs and flows of ourselves. We are not only the sum of our parts, but the squishy, emotional and immeasurable spaces in between.

Yet throughout history, being well has always been comparative – if you are more well, you are also less mortal. The biohackers are defined by their data, but are they paying attention to the right things?

PART II

Shoulders of Giants

CHAPTER 4

Elixir Mania 2.0

I know that I am mortal by nature, and ephemeral; but when I trace at my pleasure the windings to and fro of the heavenly bodies I no longer touch earth with my feet: I stand in the presence of Zeus himself and take my fill of ambrosia, food of the gods.

Ptolemy

Emperor Qin was the first Chinese emperor. From 221 BCE, after he brought together the warring nations by defeating them one by one, he used his power to upgrade his part of the world: he established unprecedented civic systems and infrastructure, standardised the width of roads and canals to facilitate transportation, minted a single currency to jump-start the economy, and conducted a census to take stock of what and who he owned. Qin believed that if he built his dominion to be as efficient as possible, his civilisation would live forever. And he wanted to be around forever to rule it.

By all accounts, he was a brutal leader. The Grim Reaper was a constant companion; he survived at least one coup and an assassination attempt, and travelled through a network of tunnels under his palace to help him commune with immortals.

What he wanted, though, was not just to talk to them; he

wanted something that would make him like them. He studied Daoist texts that described an island called Penglai where eight mythical immortals lived, and Emperor Qin decided to find them and obtain their secrets. He spent a fortune on ill-fated journeys and wild-goose chases, and appointed a team of alchemists – the *fangshi* – to reverse-engineer the eight immortals' potions; they read spiritual texts and deciphered their symbols as potential clues to possible recipes.

Today's biohackers are modern-day explorers, ploughing our biological and chemical undergrowth for a silver bullet to solve our mortal problems through data. Their adventuring follows the furrows of millennia of experimentation, from mystical elixirs first mixed together by ancient alchemists to more modern pharmacological compounds mixed together by lab techs at pharmaceutical companies.

The common elixirs of Qin's era were distilled from metals and minerals that were believed to have mystical properties, such as sulphur, gold, lead, arsenic and cinnabar (which was used to make mercury). It is likely that these were brewed and served to the emperor, who took them regularly. But none of these things are supposed to be ingested. They're poisonous.

If you drink mercury, for example, your heart rate intensifies and your cheeks flush. According to records kept by the *fangshi*, after a while you lose weight and your sexual appetite increases. But after *this*, ulcers explode across your chest, various body parts swell, there's excruciating pain, vomiting and ultimately an agonising death.

So why would you do this? Well, over the 2,000 years from Emperor Qin (who died aged forty-nine) to the Yongzheng Emperor, who in 1735 was last Chinese emperor believed to have died of elixir poisoning, there were many excuses: the side effects were explained away as the body cleansing itself of impurities.

Or, when they died, it was decided that these mortals weren't virtuous enough to live forever. Most of the time, though, alchemists were blamed: they didn't follow the recipe right, or they were incompetent, or they were fools. Many, many more alchemists died during this period than emperors, whether testing their brews on themselves first, or by execution after an emperor died.

When Yongzheng died, most of Europe had also spent centuries drinking immortality potions. *Aurum potabile*, or drinkable gold, was popular: it was considered the cure for a wide range of illnesses,[1] and was taken – and prescribed – if you wanted to restore yourself to former splendour.

But from the Middle Ages to around the seventeenth century, the most sought-after substance for immortality-seekers was the philosopher's stone. Legend has it that it could turn base metals like mercury into gold and silver, and cure disease, restore youth, and keep someone who had it alive forever. Searching for it became a popular obsession for alchemists around the globe during the medieval period and the Renaissance, and the inspiration for many great inventions and technologies – from lab techniques later adopted by medicine and chemistry, to tools used to purify and alloy metals.[2]

The alchemists' race to find the stone was not a poor man's pursuit. The equipment alone required access to both trades and secrets only available through extensive study. The materials were precious metals and other highly valued substances. Monied families were the greatest benefactors, along with monarchs and popes, such as King Henry VI, who was looking for health remedies as well as a way to finance his army,[3] and Pope John XXII, despite his edicts to ban the practice of alchemy.[4] Every apparent breakthrough first benefited the elite before it trickled down to everyone else, often brought to market by hucksters selling promises of the same remedies given to the aristocracy but derived using questionable ingredients.

As the Enlightenment began to take hold, scientific reason replaced alchemy, and the philosopher's stone was rationalised into myth and legend. But the romance of a cure-all substance persists today, and these dreams are being funded by today's wealthiest: tech billionaires who want to be the first to defy death and stay forever young.

Alchemy's mythos lives on in today's wellness industry, where the imprimatur of science is a selling point. On the border of Venice Beach and Santa Monica, not far from Dave Asprey's flagship Bulletproof café, where you can get a coffee with butter in it and packets of powders that claim to rejuvenate your mitochondria, there's a juice bar that specialises in 'plant-sourced alchemy'. It is a tiny little shop of blonde wood and blonde women, with native Californian plants lining the pathway to the inside. I've been a few times to buy their cold-pressed juices made from veggies, fruit, mushrooms and seaweed, and my wallet has felt significantly lighter every time I leave. I have never felt flush enough to add potable magnesium mixed with L-theanine 'to promote alpha wave activity' ($44) to my purchase, or a shot of medium-chain triglycerides 'that supports body and mind energy' ($29.96), but when I win the lottery, I might splurge. Because even I dabble in personal alchemy on occasion, just to pretend that I too can be as pure as the goddesses who work there appear to be.

Approximately three-quarters of Americans take some kind of supplement.[5] Roughly two-thirds of people in the UK do too,[6] and all the life extensionists take fistfuls of the things. Ray Kurzweil, who takes hundreds a day, said in 2005, 'In my view, we are not another animal, subject to nature's whim.'[7] We have agency over our health and wellness, and to Kurzweil this is our moral imperative. People take supplements for many reasons, but 'reducing the risk of age-related disorders' is one of the top three drivers of consumption around the world.[8] In 2025, the market was estimated to be worth approximately $485 billion.[9]

But unlike medications, supplements are loosely regulated because they are 'culturally embedded', which means they don't have to actually do the things they're advertised for. In the US, they were exempted by the Food and Drug Administration (FDA) when the Dietary Supplement Health and Education Act was passed in 1994. In China, they're lightly regulated as health foods, and are listed along with their purported benefits, 'including boosting immunity, antioxidant activity, memory enhancement, reducing eye fatigue, improving sleep, facilitating digestion, etc.' In the EU, they fall under food safety directives.[10] That means anyone can claim something will 'promote alpha wave activity' or 'support body and mind energy', because they don't have to be clinically tested. But the illusion of scientific evidence is an excellent marketing tool. And also – benefit of the doubt – sometimes we just don't understand why a treatment works, *yet*.

Maybe someone discovers that a drug for one thing appears to do something for another condition, and so biohackers and juice bar recipe inventors fold it into their protocols, just in case. 'Oftentimes in medicine, you have stuff being tried out off-label first', explains Arthur Daemmrich, director of the Arizona State University Consortium for Science, Policy and Outcomes (CSPO). He gives me a modern example: GLP-1, the weight-loss drugs known by brand names Wegovy and Ozempic, among others. 'Before it got approved for weight loss, people were taking the diabetes stuff off-label. And then you get enough evidence from that, then sometimes the trials follow.' Off-label experimentation can be dangerous, Daemmrich cautions, but it's also a huge opportunity.

The history of elixirs to keep us alive and supplements to keep us young is characterised by a process of trial and error, and it's not unusual in science for researchers to have only a vague idea of what's really happening behind their observations. Often, we are willing to use something having seen the evidence of an effect, despite not yet understanding why it works.

The fact that not everything is mapped out is what keeps us looking. And every time there's a technological leap, it opens up a whole new part of the map – where it might just be possible to see the thing that's been lurking in the shadows: the key to eternal life. And there's a formidable group of scientists more cautious in their outlook than Aubrey de Grey who are pretty sure they know where to look.

CHAPTER 5

The Fountain of You(th)

We begin to die as soon as we are born, and the end is linked to the
beginning.

Bret Harte

'What was the basics of ageing research when I joined the field?'
geneticist Felipe Sierra says to me when I interview him in 2024.
'It was descriptive ... And everything looked bad ... You're
declining on this, you're declining on this, you're declining on
that, and you're declining on that as well.' In the 1980s, when he
did his first study into ageing at a lab in Switzerland, his field
was literally a dead-end gig.

'Ageing was so backwaters that nobody had applied basic
principles of molecular biology yet,' Sierra explains. That
decade, a genetic technology and automation boom had made
it possible to identify which genes cause which diseases, as well
as advancing genomic sequencing and viral and bacterial detec-
tion. This was the first golden age of genetics, and a rush in
interest and funding in the field of molecular biology made it a
crowded market, so Sierra was on the prowl for something that
would make a difference.

'I did something very simple,' he tells me. 'I looked for a gene
that changes with age.' Instead of looking at something that 'goes

down', though, he flipped it. 'I looked for something that goes up.' And he found it.

Generally with genes, we think of what we inherit from our parents, like blue eyes or detached earlobes. But genes are in fact instruction manuals for making proteins, which dictate the structure and function of each cell. In two concurrent studies of the mRNA in the livers of young and old rats, Sierra found that a polypeptide (the precursor to proteins) called T-kininogen increased in production during acute inflammation caused by injury but also with age.

Around the same time, in 1992, another researcher, Cynthia Kenyon, a molecular biologist at the University of California, San Francisco, also made a landmark discovery. She and her doctorate student Ramon Tabtiang imagined there was some kind of universal mechanism for ageing: something inside us that was 'controlled by regulatory genes whose activities could be dialled up or down to lengthen or shorten lifespan.'[1] So rather than ageing being a genetic trait, it could happen because certain genes switched things on or off. If that was the case, it could be possible to adjust the whole ageing process. The key to the fountain of youth was actually inside us all along.

Kenyon decided to test this idea out by experimenting on a tiny roundworm nematode, *Caenorhabditis elegans*, or *C. elegans*. Normally, their lifespan was only a few weeks, but if they didn't eat much and if they had been genetically modified not to get to reproducing age, they lived longer. Why was this, wondered Kenyon. 'I sat there, feeling a little sorry for them, and then wondered whether there were genes that controlled ageing and how one might find them.'

Kenyon and Tabtiang edited one of the worm's genes: *daf-2*. In a time period of firsts, it was the first time anyone had looked at *daf-2* specifically in ageing, and, remarkably, the worms that had been tweaked lived more than twice as long in adulthood as their non-mutant counterparts. The pair did it again and again: '*daf-2* mutants were the most amazing things I had ever seen,' Kenyon wrote. 'They were active and healthy and they lived more than twice as long as normal. It seemed magical but also a little creepy: they should have been dead, but there they were, moving around.'[2] This discovery set molecular biologists on fire – the real golden ticket would be to find something like this gene in humans.

Just five years later, a research team at the Ruvkun Laboratory at Harvard Medical School sequenced *daf-2*. In the paper they published in *Science*, they described it as homologous to human insulin and insulin-like growth factor receptors.[3] IGF-1 regulates metabolism, growth, and reproduction. It's one of the hormones that peaks in production during puberty, and falls off a cliff around thirty.

'This was a stunning finding,' wrote Kenyon.[4] 'Hormones, evolutionarily conserved hormones, controlled ageing.'

'With a single gene she changes everything,' Sierra recounts. Getting old was no longer an inevitability. 'Once that started happening, the rest of the world started taking us more seriously.'

∞

Imagine 'old age'. What comes to mind? For many, it means losing faculties, and losing function. Would you want to extend your life if the extra years meant living less?

Ageing is something we all experience, at different rates and in different ways. In 2015, in their *World Report on Ageing and Health*, the World Health Organization defined healthy ageing as 'the process of developing and maintaining the functional ability that maintains well-being in older age.' Key to this definition is functional ability – 'the health-related attributes that enable people to be and to do what they have reason to value.'[5] Sierra describes this to me in the context of 'intrinsic capacity': 'This takes into account that what you want to do when you're eighty is not what you want to do when you're twenty. So it's not that you have to have the same capacity when you're eighty as when you're twenty or thirty.'

Whatever that is depends on your interests, of course: if you want to go skydiving at eighty, but can't get on the plane because it's become more difficult to get upstairs, your intrinsic capacity has declined. If you're eighty and you have no physical trouble getting on the plane, but you have cataracts and you'd rather read than parachute, your intrinsic capacity has also declined. So it's not just the presence, relative encroachment, or absence of a disease, or even the absolute number of years of life that's important, but the quality and the quantity of those years in the context of a range of other factors, such as lifestyle. This is qualitative, and so difficult to measure objectively. The term used for it in age research is 'healthspan' – and the objective in most contemporary science in this field is to increase it – the amount of healthy life, rather than the number of years lived. Healthspan means staying healthy for longer, and keeping the diseases of old age at bay. It means keeping your body young, intervening in the process of growing old, and considering the possibility of 'rejuvenation.'

Scientists who work in this space have found themselves on the edge of respectability – the fringe of the fringe, a doctor colleague said to me dismissively when I mentioned I was researching the topic for the *Immortals* podcast. Kenyon's discoveries brought the field of age research closer to understanding that the controls to rejuvenate ourselves might actually be within us, and that maybe these can be manipulated to modify the ageing process.

∞

Scientific knowledge is more like a process than a single moment in time. There is no endpoint, but a forever path ahead, littered with old, disproven ideas that, perhaps, were actually the stepping stones that got us to where we are now. Scientific breakthroughs are usually the beginning of a new fork in the road that opens up new lines of research and the opportunity for discovery. And long before Kenyon or Sierra found genetic links between age and hormones, two other notable scientists were also at a fork in the road, living in an era of scientific discovery that has led us to the place in longevity research we are now. It's just that they never knew what they were looking for.

The first story begins with an improbable announcement made by a distinguished and celebrated physiologist and member of the Royal Society. It was 1889, in Paris, when Charles-Édouard Brown-Séquard stood before an esteemed audience of peers at the Société de Biologie and gave what became his most famous presentation. Sensing himself growing old at the age of seventy-two, Brown-Séquard had started investigating loss of sexual function, and claimed to have found the right treatment: an injection into both of his arms of a mixture of blood taken from the gonads, semen and 'juice extracted from a testicle'[6] of a dog or a guinea pig.[7]

Stunned silence.

Before you reject this as quackery, Brown-Séquard's connection between the testicles and the fountain of youth was rational. There had been research suggesting that something as yet unidentified was a salve for the effects of age (at least in men), but his conclusion was a little off. Despite his own reported improved energy levels, muscle rejuvenation, cognitive functioning, and heightened abilities and skills in the bedroom, by 1893 none of the formal experiments underway, nor informal, off-label prescriptions issued by many physicians, found any conclusive or repeatable effects for rejuvenation, or, indeed, any of the disorders of 'age' they were intended to treat. But the connection he made would lead to something useful. Eventually.[8]

The second rejuvenation treatment was the brainchild of one of Brown-Séquard's apprentices, Serge 'Samuel' Voronoff. A promising pupil, he was a student under the eugenicist Alexis Carrel, a Nobel laureate and pioneer of organ transplantation.

Voronoff took his mentors' two novel ideas – putting an organ from one body into another, and the mysterious rejuvenating power of testicles – to develop his own theories about vitality. He was working in the field in Egypt with eunuchs and he came to believe that the absence of testicles was responsible for their ageing. And so, rather than inject semen into himself, he transplanted sex glands from the young onto the old – in order for the old to become young again.[9]

He first tried his idea out on rams. The evidence pointed to positive results, with the animals demonstrating more vigour. One reportedly began to sire offspring again. Then Voronoff moved to grafting thin strips of testicle skin a few centimetres wide from recently executed prisoners onto older men's scrotums. Again, the positive results: the patient's fatigue and physical symptoms of age seemed to reverse. But after word got out, Voronoff became a victim of his own success: the procedure became so popular that the supply couldn't meet demand. Voronoff turned his scalpel to young monkeys.

In 1923, he announced the rejuvenation properties of the sex glands to scientists at the International Congress of Surgeons in London, informing them that the procedure had been so successful that the Pasteur Institute had built him a chimpanzee-breeding compound in West Africa.

Over the next decade, Voronoff personally transplanted chimpanzee and baboon testis fragments onto forty men – from millionaires to labourers, university professors to painters. Most were over the age of forty. The oldest was eighty. He published books and papers, and was quoted as saying life could span more than 140 years;[10] the press described the outcomes of his procedure as miraculous and life-changing. Rumours swirled around about wealthy men paying to get to the front of the line. Over a period of twenty years, more than forty-five surgeons around the world used his technique more than 2,000 times; 500 men were operated on in France alone.[11] In Brazil, where he was a minor celebrity, Voronoff inspired Carnival songs.[12]

In 1926, he transplanted monkey ovaries into women to overcome the consequences of menopause, and a human ovary into Nora, a chimpanzee. We have no record of what happened to the women, but Nora appeared to recover from her reproductive sunset: *Time* magazine reported that she stopped her period for ninety days in an 'apparent pregnancy'. Sadly, it didn't come to term. 'No Ape-Child,' ran the headline.[13]

Ultimately, Brown-Séquard and Voronoff appeared to be on to something that seemed to affect men's ageing, they were wrong about the active agent: it was the hormone testosterone. And when this was isolated in 1935 by Ernst Laqueur and then synthesised a year later, making it readily available at quantity, their debatable procedures were rendered obsolete.

∞

In the early 1990s, Mike Conboy was working at the lab at Harvard when one of his colleagues dropped by for a favour. An old school friend of hers, named Irina, was coming to the US from the former Soviet Union for a couple of weeks. The Berlin Wall had fallen, and Irina and her family were finally allowed to visit.

'She showed my photograph to every single guy at Harvard lab,' recalls Irina, 'and she said, "This is my beautiful friend Irina. I don't have time to entertain her, but can you take her out to the movies or go dancing?"'

Irina arrived in Massachusetts to a full social calendar.

Mike took Thursday.

Thursday night was Latin night at the legendary Narcissus club.

'So I dressed up suave' – Mike ironically pronounces it 'swah-vay' – 'and I picked you up.'

'And I dance very well,' Irina says.

Mike chuckles. And the rest is history.

When I meet them in their office in the bioengineering department at the University of California, Berkeley, the Conboys are wearing matching tie-dye sweatshirts, and finishing each other's sentences. They are clearly as in love today as they were when they hit the dance floor thirty-five years ago.

They are also charming, warm, and a bit shambolic. Their tiny white pup is comfortably chewing on his leg on a saggy sofa, next to a sheaf of papers. Mike sets several grocery bags of baguettes and crisps down on the table in the centre of the room, where they dream up their ideas about rejuvenation.

When Irina and Mike were doing their PhDs in cellular immunology and cell biology, respectively, at Stanford University, Cynthia Kenyon came and gave a talk. 'She looked like a super young, very hip professor,' says Irina. 'And she suggested that simply by changing the intensity of certain molecules, you can make an old animal younger. Now, that animal was a very, very small, simple animal,' she says, but that gave them an idea: what if the universal mechanism demonstrated in Kenyon's *C. elegans* research could be applied to mammals?

After Kenyon's lecture, the pair started tweaking molecules in *Drosophila*, focusing on a particular one called a node receptor. By manipulating the node receptor, they found they could regenerate a fruit fly's old muscle, effectively making it young again. It was their first major breakthrough.

They wanted to go bigger – not just look at a single tissue, but the whole animal. Fruit flies, mice and humans share similar genes for coding muscle. This is why the fruit fly is a useful model organism for studying muscle biology in mammals. 'Muscle is replaced by scar tissue – that is the same defect people have,' says Irina. They tried the fruit fly study on a mouse, commonly the next organism in the laboratory hierarchy towards humans, and voilà – the old muscle seemed to go back in time. 'Suddenly,' she says, 'we had this intervention – activation on command – and that rejuvenated old muscle.'

'So we had this question: why is it that all the tissues of the body seem to kind of grow old together?' asks Mike. It doesn't matter whether they're on the outside or on the inside, whether they're exercised or going along for the ride. 'Everything seems to go to heck in a handbasket with age,' he says. They knew about the trigger, and they wondered if there was some kind of signal in the body that changes the molecular structure of muscles, and ages them all simultaneously. They set out to find out what all tissues have in common.

Think back to your biology class in school, and you might remember those hand-drawn anatomy posters with elaborate illustrations and labels for each part of the body. I have three in my bathroom downstairs: the skeletal system – the body's framework of bones, ligaments, cartilage and joints that gives us structure and shape; the nervous system – the network of nerves that transmits electrical signals between our brain and the rest of our body; and the vascular system – the blood vessel pipelines that take nutrients and oxygen to our muscles.

Could it be that the 'heck in a handbasket' alarm was passed through the body via one of these systems?

'The ideal experiment would be: "Okay, what if we transplanted a young nervous system into an old mouse,"' explains Mike. 'Well, we can't technically do that. What if we put a young vascular system in an old mouse? We can't do that either.'

Irina looks at me with a twinkle in her eye while Mike continues: 'What if we put young blood into an old mouse?'

This would have to be more than a one-off injection – they would need to transfer a large proportion of blood from a young mouse to an old one to make sure there was an observable effect. At that time, in the early 2000s, there was no technology to give a mouse a blood transfusion. Inspired by two recent papers out of Stanford on a different topic, Mike says, they had an idea. 'What if we connected the young and old mouse together? You can take an old mouse and stitch it to a young mouse, and instead of the skin healing edge to edge of the old mouse, it heals *across* edge to edge to the young mouse,' he explains. 'As that tissue heals, blood vessels re-form there. Now you have blood slowly trickling from one animal to the other and back and forth.'

This is called parabiosis.

There is evidence that there were animal grafting experiments stretching back to ancient and medieval times, but French zoologist Paul Bert, in 1864, was the first to postulate what would happen if one was to connect two creatures together. Bert demonstrated that the procedure created a single circulatory system between two mice, by injecting fluid into one mouse and watching it pass into the vein of the other.

For this, he received an award from the French Academy of Sciences for innovations in experimental physiology. Thanks to further experiments by other researchers, stitching two mice together, we have learned that tooth decay is caused by sugars in the mouth, not in the blood,[14] that immunity may pass through blood and can tackle specific cancerous tumours,[15] and that problems with the hormone leptin can cause obesity.[16] Mike and Irina's 'heterochronic' study was the first to look specifically at tissue regeneration.

After some preliminary research in petri dishes, they sutured three categories of mice together: young and old, young and young, and old and old. All the young partner mice had a modification that made proteins in their blood shine green under a UV light. That way, the researchers could tell which cells belonged to them. They let the blood flow for five weeks, and then gave each mouse an injury in its hind leg muscle. Five days later, the mice with the young blood – including the old ones – recovered 'robustly', unlike the un-tampered-with oldies, whose muscles regenerated poorly, 'typical of aged animals'.

The team tried again elsewhere, on other body parts. The liver, same thing. The brain, same thing. As the results began to trickle in over the lifespan of their conjoined rodents, the Conboys were able to conclusively say that the tissues of the old mice looked more like the tissues of a young mouse – whether they were in the muscles or in the liver or in the brain. Something in the young blood rejuvenated the old tissues.[17]

Laboratories all over the world followed with their own results, but far more testing was needed before anyone knew what was really going on to rejuvenate the tissues, and before young blood could be used to treat old humans.

∞

By this time, Sierra was monitoring all the strands of research – from plasma to genes, from flies to mice – from his position at the National Institute on Aging (NIA) in Bethesda, Maryland, as director of the Division of Aging Biology. He figured that, as a person sitting in a centre of scientific credibility, he could bring the threads together to support his central mission. On his watch, ageing would no longer be about decline; he would make it the star of the show. And, he'd also be able to draw a line in the sand between the establishment academics like him and those like Aubrey de Grey who promised radical life extension.

Sierra recruited his deputy director, Ron Kohanski, to help him develop a compelling pitch for a new transdisciplinary branch of ageing research. 'Through several coffee-soaked discussions at the Starbucks across the street from our office, we refined the ideas and arguments,' he later wrote in the *Cold Spring Harbor Perspectives in Medicine*. It would be called 'geroscience.'[18]

First they won over the director of the NIA, but then Sierra needed to convince the other twenty-six institutes and centres that form the National Institutes of Health (NIH) to collaborate across their specialisms. He called a meeting, expecting a pushback. That didn't happen.

'Tony Fauci could hardly stay on his chair,' Sierra laughs, remembering the response of the then head of the National Institute of Allergy and Infectious Diseases. 'He said, "This is fantastic – what are we going to do now?"'

Sierra organised the very first global geroscience symposium in July 2013. Five hundred scientists and advocates attended; the NIH's then director, Francis Collins, the geneticist who had led the Human Genome Project, gave the keynote. For Sierra and the other scientists, it was a massive political win, setting the agenda for a brand-new, extended, healthier future, and placing themselves apart from de Grey and the immortalists. 'Aging biology has reached a tipping point for research,' Sierra tells me. They established a research 'roadmap' that prioritised biological processes that seem to accelerate physical decline: inflammation, immunity, adaptation to stress, epigenetics, metabolism, macromolecular damage, proteostasis and senescence. Long on the outside looking in, attendees at last felt their ideas about the biology of ageing were part of the establishment. 'It rallied the field,' Sierra tells me. 'And a lot of people from outside of the field said, "Hey, this is interesting."'

This was the same moment Calico entered into the mix, and people who weren't geroscientists began to take an interest too – the longevity entrepreneurs.

∞

There's a freezer in the Conboys' research lab at Berkeley. It takes up a quarter of the room and inside are hundreds of frozen mice who've had their blood doctored.

'Are you surprised the longevity community has jumped on your research?' I ask Irina.

She responds immediately. 'Actually, they always jump on our research, which is pretty cool.'

Mike is less enthusiastic. 'I worry sometimes when I see more of the . . . lay public thinking about experimenting on themselves,' he says, choosing his words carefully. I've just been telling him and Irina about my conversation with Bryan Johnson and my research into biohacking. 'I realise every day how little I know.'

I'm not the first journalist who's spoken with the Conboys this week. Their phone is forever ringing with people looking for a quote, as well as people looking for treatment, or – increasingly – curious entrepreneurs.

'Many make decisions about what works and what doesn't work based on the title of the paper. They do not understand the data,' says Irina.

Mike looks a little deflated. 'I can't control how someone interprets my research, especially if I suspect that they don't really understand it or read it properly. All I can do is publish as clearly as I can.'

'If you think that we already have it and you can simply go and do the procedure and become younger – not yet,' Irina says. 'If the goal is to stop ageing and plateau, or start gradually becoming younger, I would just say that, right now, we don't have it. But if you invest correctly, your time and efforts and resources, we will have it soon. That is my feeling about it. Soon.'

'That's what's traded in Silicon Valley,' says Mike, ruefully. 'The potential.'

I had first heard about the Conboys' research back in 2017, when investigating the unusual longevity promises of a start-up called Ambrosia Health.

Jesse Karmazin, a Stanford Medical School graduate, had opened Ambrosia Health's doors in a business park near a redwood forest in Monterey, California, to run an experiment testing the Conboys' hypothesis that something in young blood regenerated cells. There still wasn't much to go on about *why* young blood appeared to reverse ageing – in the decade of research since their parabiosis papers, only a few molecular and hormonal candidates had surfaced, but even in these the evidence for them was inconclusive, and all of it was in mice.

There were two human trials underway at university departments to find out more – one for people with advanced dementia and one for people of advanced age. The Conboys had published a paper in *Nature* in 2016 that hypothesised that young plasma transfusions might be just as effective as parabiosis in rejuvenating mammal tissues. Karmazin read it during his psychiatry residency at a Harvard hospital after completing his medical degree at Stanford, and he saw a gap in the market: he decided to run a human trial of his own as a private company not associated with a university – to test whether transfusing young blood plasma would reverse ageing in anyone over the age of thirty-five.

Human trials are expensive, so Karmazin decided to charge people who wanted to participate. When I spoke with researchers who have designed and overseen clinical trials for many public health interventions – from AIDS to COVID-19 – they expressed surprise and concern that he was charging, and so much: the entry fee to participate – to have biomarkers measured, to receive a transfusion of one litre of young plasma from a donor between the ages of sixteen and twenty-five years old, and to have their biomarkers measured again for comparison – cost $8,000. What effect could the participants' investment have on their subjective feelings of wellbeing after they received the experimental treatment?

His research proposal was called 'Young Donor Plasma Transfusion and Age-Related Biomarkers'. He phoned Irina Conboy to see what she thought; Irina declined involvement. She didn't agree with the premise, she tells me. Karmazin's study didn't 'fit very well', she says. There wasn't enough evidence to look for positive effects so broadly. But Karmazin ploughed ahead anyway. He launched the study onto the main US clinical trials search tool, ClinicalTrials.gov,* and then he went to Silicon Valley on a recruitment drive.[19]

The Valley should have been a good fit for Ambrosia Health. By this point, Peter Thiel had spent millions of dollars supporting longevity research by investing in the Methuselah Foundation; and Sam Altman had also been actively looking for investment opportunities in plasma therapeutics.

This trend was more than common knowledge – it was a cultural trope: in the fifth episode of season four of the hit HBO TV show *Silicon Valley*, which first aired in 2017, tech entrepreneur Richard is about to give a presentation to the idiotic and power-hungry CEO Gavin Belson. As he begins, a young man comes into the room wheeling a medical device. Richard stops speaking and watches as the young man connects a needle to his arm, and then another to Gavin's. Without a word, he switches the machine on and settles into a chair next to Gavin. Richard, a little surprised by the interruption, asks, 'Uh, is this your assistant?' Without missing a beat, Gavin responds, 'No, of course not. He's my transfusion associate . . . Regular transfusions of the blood of a younger, physically fit donor can significantly retard the aging process. And Bryce is a picture of health. Just look at him. He looks like a Nazi propaganda poster.'

Karmazin's pitch was similarly tight. 'In Greece, there was something called Ambrosia,' he told a longevity podcast in 2018. 'It was basically the food of the gods that gave them immortal life, but also good health. And, you know, in mice that's what young blood seems to do. And so that's why I named the company Ambrosia.'[20]

Jesselyn Cook was a tech reporter at the *Huffington Post* at the time. 'It was definitely a big story, and I think it just had all the

---

* It is clearly stated on every page that the FDA does not look at or check any of the clinical trials that are listed there. However, this registry is another one of those scientific kitemarks that makes a study look official.

elements of virality, she recalls when I speak with her for *The Immortals* about her own two-year investigation into Ambrosia Health. 'You have teenage blood, you have vampires, you have age reversal, you have elements of Silicon Valley greed.

'Karmazin was quoted saying things like, "I wanna be clear at this point: it works, it reverses ageing", and "I'm not really in the camp of saying this will provide immortality, but I think it comes pretty close",' she says. He called it 'plastic surgery from the inside out' to *The Times*.[21] In *Rolling Stone*, he was quoted saying 'It makes people younger'[22] The company and the trial got more than one hundred press mentions in under two years. 'It doesn't matter if you were young, wanting to feel extra capable, or if you were old and trying to, in his own words, reverse your age, Cook tells me. 'It just seemed like no matter what the problem was, Ambrosia could help.' Thiel got entangled in the flurry of press as rumours swirled that he was a client. He was not. Soon, Karmazin was appearing at Re:CODE, a thought-leader conference, where Hillary Clinton, then California Senator Kamala Harris, and Reid Hoffman, the CEO of Netflix, were also participating.

And though Karmazin was bullish on stage, Cook found him more reserved – if not avoidant – when asked directly for the research to back up his claims. Every time she spoke with Karmazin, he said he'd back them up 'soon.' She – and many scientists – weren't convinced that he wasn't cherry-picking his results for profit, or compromising the quality of the study with a placebo effect. Karmazin did not have a licence to practise; a doctor he brought in to run one of his clinics told Cook he'd not seen the data either. They needed the evidence that he wasn't misrepresenting a supplement by dressing it up in medical clothing and requiring a qualified medical professional to administer it. The Conboys had lost control of their research findings to the hype and thrill of Silicon Valley. But Ambrosia Health was never able to get VC funding. The recent meltdown of Theranos, the disgraced start-up that promised miraculous diagnoses with a single drop of blood, had turned many VCs wary, one funder told me.

Young plasma treatments for the physical and cognitive effects of age had never been approved by the FDA, and yet by 2019 Ambrosia Health had a total of five clinics around the US, and Karmazin was hinting he was looking to open another in Manhattan. None of the researchers or bioethicists I spoke with about Ambrosia Health trusted Karmazin's experiment and they were deeply concerned that other private clinics, also claiming to do research studies, had also popped up, offering paying clients young blood transfusions.

The procedure was being marketed as a medical treatment that had never been proven, and so it's unsurprising that the FDA stepped in. On 19 February 2019, they released a statement:

> The FDA has recently become aware of reports of estab-
> lishments in several states that are offering infusions of
> plasma from young donors to purportedly treat the effects
> of a variety of conditions ... Simply put, we're concerned
> that some patients are being preyed upon by unscrupulous
> actors touting treatments of plasma from young donors as
> cures and remedies. Such treatments have no proven clinical
> benefits for the uses for which these clinics are advertising
> them and are potentially harmful.[23]

Ambrosia Health went dark within hours of the statement. When I exchanged emails with Karmazin in 2023, he told me, 'The FDA never reviewed our data before making their announcement.' He continued: 'I have no idea if they were ever aware of Ambrosia, in fact. They have never mentioned

Ambrosia. It is entirely possible they were referring to our competitors, who acted with less diligence and respect for the rules.'

∞

Throughout the twentieth century, technology supported science in the pursuit of knowledge. As the two fields intersected, their practitioners' confidence in the idea of them working in tandem to find medical treatments only increased. Today, there are hundreds of start-ups, foundations and non-profits trying to maximise the flood of knowledge about the process of ageing. It's foolish to think this new crop of investors aren't also looking for big financial returns, now.

'They want results today. They don't want results ten years from now,' Sierra explains. It's not that they're against geroscience; it's that they don't like how slow and methodological it is. 'Real science does not work that way.' One researcher at a private start-up looking into plasma-based treatments to target ageing who I spoke with, who didn't want to be named, told me that she felt pressure from the 'money people' to get results. 'At times we did feel we had to speed up,' she told me. 'Start-ups are high pressure.' Yet Sierra knows that every time a promise isn't fulfilled, the field loses legitimacy. Still, a story about the miraculous properties of young blood or a headline about today's 60-year-olds living to 1,000 will get more clicks than a research study about a molecular tweak in a nematode.

The two sides need each other: the entrepreneurs and the optimists need the scientists to give them legitimacy, and the scientists need the headlines that come from big promises to bring their field into view. 'One of the first people who came out of the woods to congratulate me on the concept of geroscience was Aubrey de Grey,' Sierra says. 'I remember clearly when it happened. I was at an age meeting, and I was talking with some people, and Aubrey was talking with other people, and all of a sudden he sees me, he breaks from his group, walks straight to me, and says, "Felipe, that was brilliant."

'Scientists are extremely careful, and they don't want to push too hard,' Sierra continues. 'Basic scientists . . . just want to have fun in the lab and understand how things work. That's why we spend so much time doing so much work on a very small thing.'

So is there truth to de Grey's criticism, that biologists don't want to let things out of the lab?

'You're not going to go from eighty to 120 with one drug or one treatment,' he cautions. 'It's going to be slower than that. There could be a big breakthrough one day, but it's not going to be to immortality, that's for sure. It's one thing to increase the lifespan of *C. elegans* by 40 per cent. Then you try it in mice, and it's maybe 15 per cent. Well, maybe when you try it in humans, it's gonna be 3 per cent, which is already fantastic! But it's not gonna be 40 per cent! There's no way . . . I don't want to be too pessimistic and say, no, we're not going to get there . . . but it's going to take much more effort and time than other people think.'

De Grey disagrees. 'People are too worried about their grant applications being canned by someone who wants to call them irresponsible, whether or not he actually is irresponsible,' he says. He's not one to shy away from saying what's on his mind.

'I mean, the guy is very bright. Nobody doubts that. He's just a little bit off the main course,' says Sierra.

And so are the people de Grey is now partnering with, a group that doesn't just *want* to live forever – they *know* – in their cells – that they will.

CHAPTER 6

Conventional Wisdom

Verily, verily, I say unto you, If a man keep my saying, he shall never see ✗
death.

John 8:51

It's early October 2022, and the sun is scorching in downtown
San Diego. In the ground-floor conference room of the Town
and Country Resort, a charismatic man is vigorously pacing
back and forth across the stage in Italian loafers. He's wearing
a fashionable slim-cut suit, shirt open at the collar to reveal a
tanned chest. He has swept his movie-star chestnut hair back
from his crease-free forehead. Like Bryan Johnson, very few of
his facial muscles move when he speaks. The only giveaway of
his golden years are his eyes, which are so conspicuously aged
that he looks like a collage of a person assembled by an alien.

'I want to see you all dancing next time when this music
plays,' he chides in a surprisingly high-pitched, unaccented
American voice, 'because look – we're supposed to be the most
alive people on the planet, aren't we?'

James Strole, co-founder of RAADfest, anti-death activist, and
Executive Director of the Coalition for Radical Life Extension, is
glistening in the bright lights as the thousand-strong audience
whoop and holler at his feet on Day 1 of the three-day gala.

'We should show it! We can't be afraid of showing it! We need to really come out there with our aliveness! It's good for our immune systems!'

The room of mostly over-sixties get up from their chairs and give Strole his first standing ovation of the morning. The majority of the attendees are sagging in the correct places, but there are many faces and figures with the telltale signs of plumps, fillers and lifts. This room is full of the radical life extension-curious, here to celebrate the Revolution Against Aging and Death (RAAD). Strole, their leader, gently raises his palm. They fall silent.

'Everything we're all experiencing, human beings are experiencing, these last couple of years, seeing people that we care about and love pass. Seeing people get sick at enormous rates. Much less the environment that has not been very conducive to really being excited about living.'

The audience murmurs. They've had to work overtime to keep the faith during the COVID-19 pandemic.

'A few months ago,' Strole whispers to the room, 'I began to feel a new strength that I've never had before. There is a new spirit of energy coming out of me.'

A few Oh-Yes-es and I-See-It-s.

'We need to have a spirit of immortality, a strength in our bodies!' he commands. The crowd struggles up from their seats again.

'I feel like there is a tapping into this immortal lineage, or whatever we want to call it, in our bodies to let there be a new birthing of life!'

Strole soaks up the rapturous applause.

∞

RAADfest is billed as a global age-reversal public event. Its website declares:

> RAADfest is the largest and most immersive event in the world focused on super-longevity for a general audience. Bringing together cutting-edge science, inspiration, entertainment and fun, RAADfest is more than just a conference – it's a celebration of life. From brain longevity and sexual health, to senolytics, personalized medicine and helping your pets live longer too, RAADfest provides the information and inspiration to enable people to take charge of their longevity.

Despite its unusual subject matter, it looks like most corporate events: a line-up of luminaries, lots of congratulatory back-slapping, networking breaks, sponsor adverts, and an expo hall where you can sample the latest products to keep death at bay.

'The innovations and the therapies that are available there are extraordinary,' Strole enthuses from the stage. 'I got an IV today and I want to encourage you guys if you need a little boost to your immune system, get an IV.'

An IV of what? There's a whole menu to choose from. Plus there are countless creams and headsets and cold plunges and injections and mushroom vitamins and blood panels in every booth.

'There's a DEXA scan. That is a very, very low price for DEXA scan, by the way,' continues Strole, describing a diagnostic test that measures bone density – usually to identify osteoporosis, but which has become a popular biohacking and wellness tool to help tell what are normal age-related changes and what are not.[1] 'And there's so much more. So much more, so much more. There's so many fantastic things you can do. Yeah, I'm signed up for, like, twenty-seven things.'

I need a glossary to decipher most of the treatments. I'm starting to feel like I've missed the memo. It's a lot to assimilate, but Strole assures us: 'Don't worry about what you don't understand at the moment. The longer you're in the conversation, the more you get really clear what you need to apply for yourself.'

RAADfest is 'the Woodstock of biotech'.* By 2022, its seventh year, RAADfest attendance was averaging around 1,000 people. People came from all over the world, paying $727 for the most basic tickets and $2,997 for access-all-areas passes.

As I tuck into a complimentary plate of organic food, I thumb through an app that uses 'ancient technologies' to retrain brain waves to be resilient to modern chronic toxic stress. It's not helping. I wander past a pitch for 'bliss in a bottle' cream, which the presenter says has resolved muscular atrophy and paralysis with one application. He saw it happen himself.

Most of the plenary talks tread a version of the same path: we have been fooled into thinking that ageing is inevitable; it can be conquered with a simple solution; here's an app/cream/injection/treatment that uses the science 'they' are keeping locked up in their labs, or that the government doesn't want you to access. I can't see a single treatment on offer that's been FDA approved – and many people are seeking funding for clinical trials. Tech entrepreneurs with deep pockets are roaming the aisles looking to get in early on a treatment that could make them their next million. Everyone here is a believer in some way.

Perhaps none more so than James Russell Strole. Born around 1949, he has actively pursued a life that never ends for more than fifty years. In fact, he believes that humans don't actually need to seek

---

* Not the Glastonbury of biomedicine, nor the Coachella – this conference is catering to a Baby Boomer demographic.

immortality because we are fundamentally physically immortal. Yet somehow, as civilisation has evolved, we've forgotten that. Instead, we've found ourselves trapped in a powerful cult of death where systems are designed to sustain themselves at the expense of our own lives. This is evident in how financial systems, work environments, urban development and even how we envision the future all hinge on replacement, rather than renewal. And because it serves how the world is currently organised, he believes we have been indoctrinated into this progress towards finality, and now we imagine only the spirit will live forever, not the flesh.

Death doesn't have to be this way, says Strole and his believers. It's time to end The End.[2] If we make an effort to live well, by taking care of our bodies according to the latest scientific advances, and our minds through positive thinking, there's nothing to stop us from resurrecting our immortal lineage and living forever.

RAADfest is Strole's marketplace. It sells improbable dreams and unverified remedies – all of which are legal, he assures us. On this year's speaker list are lifestyle influencers on a longevity kick, entrepreneurs selling their unproven gene therapies, naturopaths promoting questionable injections, a roboticist using AI in elder social care, and Ray Kurzweil, the Google technologist who is convinced that the way to defeat death is to merge with AI.

Strole was nineteen and working as a real estate agent when he first heard Charles Paul Brown and Bernadeane Brown proclaim their immortality at an event in Phoenix, Arizona, in 1968. After a stint as a nightclub singer, Charles had become an Assembly of God evangelical minister and moved his sermons to the local radio station. Bernadeane was a preacher's daughter-turned-model. Both in their early twenties, the pair had discovered their mutual passion for immortality. Charles wore loud neckties and already had a following for evangelising about unlimited lifespans; Bernadeane had a severe platinum bob, killer red lipstick, and an affinity for leopard print. It was 1960. They were soon married and had two immortal children – Kevin and Kim.

Their message was a reinterpretation of the Bible – Jesus wasn't promising spiritual salvation; he had come to grant us physical salvation. We were already immortal, we just had to believe it. In 1968, James fell for the good word, Charles fell for James, and soon after, all three became professional and romantic partners in their mutual immortality journey.

The throuple hit the road as CBJ and later traded under the banner 'The Eternal Flame Foundation' (they would have a series of rebrands over the years). Throughout the 1970s, 80s and 90s, they gathered a flock of searchers who were won over by their wild idealism, bombastic presenting style, and hours-long sermons. The lectures were heavy on Bible verses, the audience often weeping, howling and fist-pumping, but light on toilet breaks or snacks. CBJ believed in 'cellular intercourse', a practice involving hugs, handholding and communicating that, according to Strole in CBJ's first book, *Together Forever: An Invitation to Physical Immortality*, 'exploded us right out of traditional mortal living' when the three of them were together.

Cellular intercourse was the best way to achieve 'cellular awakening', what they described as a vibration in the sleeping immortal cells in our bodies upon realising they (the cells) were not born to die. It's a full-body and full-psyche meltdown – similar to the prostrations of mania evangelical Christians experience during fits of religious ecstasy. The cellular awakening Jeanne Marie Laskas witnessed at a CBJ gathering in Tel Aviv and documented in 1991 happened to 'a curly-haired Israeli woman in a state of near hysteria. She was handed a microphone . . . "I feel!" she shouted finally. "I feel! I don't know what I feel. It's something I've never been through, my body is making the choice, my body is talking to me, my body is saying so many things!"[3] It is cellular awakening, CBJ promised, that releases you from the cult of death, and brings you into the cult of life.

The organisation was named a 'new religious movement', a term used in sociological research to avoid the pejorative term 'cult'.[4] In 1991, CBJ – by then with 30,000 people from around the world on their mailing list – earned a place in an archive of religions and cults pulled together by Rick Ross, who had worked as an educator and deprogrammer since the mid-1980s. CBJ had received donations and membership fees and sold tickets which, by one estimation, brought in around $1.4 million every year.[5] 'I want you to see how important it is to lay some money down on the three of us!' Bernadeane would say in their sermons.

But as CBJ began to expand in size and wealth, the Internal Revenue Service became curious about how the trio could afford to live in a luxury pink villa with a pool and a jacuzzi in North Scottsdale, Arizona, and drive a Cadillac and a Harley-Davidson, when they were supposed to be a charitable non-profit. The accounts weren't transparent, money was misplaced, and the group lost several lawsuits. Between 1994 and 1995, the throuple began to separate, Rick Ross republished rumours of group sex and forced homosexual relationships – though no one was ever charged – and one of Eternal Flame's high-profile members died.[6] All of this destabilised their members' confidence. There was an eye-opening exposé on the US current affairs TV show *60 Minutes* that alleged bad behaviour and brainwashing. Charles left, but miraculously the Eternal Flame didn't die. It did, however, evolve.

James and Bernadeane changed the name in the mid-1990s to People Forever, and adjusted the doctrine to be more in line with other New Age religions: less Jesus Christ and more metaphysical spice. 'Scientism' made its way in, and People Forever became an alternative health movement, rather than a religious organisation. They consolidated their efforts around Scottsdale's hub for alternative therapies, incorporated as a business in 1996 under yet another new name, People Unlimited, and built a core group of around 120 local regulars, plus a more expansive global community of practitioners, researchers and immortalists. Charles returned to the fold some time later, humbled and repentant, and People Unlimited claimed they were the largest immortality network in the world.

It was another discordant moment for the group when Charles died of complications associated with Parkinson's and heart disease in 2014 at the age of seventy-nine; their founding father was not immortal after all.[7] As one might expect, they lost another crop of followers. Bernadeane's new beau, Joe – People Unlimited's communications director – became responsible for fielding the schadenfreude about Charles, and explaining it away: he didn't live an adequately pure life. The organisation rapidly moved on without glancing back. Today, People Unlimited host events in Scottsdale, and oversee a website with relevant immortalism resources and an 'Ageless Education Program', which includes a contribution from de Grey. James and Bernadeane dropped Charles's doctrine of cellular intercourse.*

I ask de Grey about his experience working with People Unlimited after Charles died. 'It was painful,' he says.

People Unlimited is an interesting organisation, I respond. De Grey smiles ruefully. 'You are being very euphemistic,' he chides. 'They're a disaster area.'

So why partner with RAADfest, I probe.

---

* Bernadeane wasn't able to attend RAADfest 2022; she was at home with COVID.

With RAADfest, he explains, James and Bernadeane wanted to start something 'that had a bit more science in it', and that was what piqued his interest in the first place. He's spoken at every event since its inception. He's watched the organisation evolve from the inside. It hasn't been an easy transition; 'they had the albatross of the reputation of People Unlimited', he says. At the first conference, in 2016, he was one of only three credentialed people who spoke. 'A lot of happy clappy nonsense', he recollects. 'But it wasn't *all* happy clappy nonsense. So I thought, you know, there's potential here. And I carried on going. And speaking of course.'

Over the eight subsequent events, de Grey came to the conclusion that step by step it was coming together. 'I wrote an editorial in my journal *Rejuvenation Research* about six years ago in which I talked about how, you know, they pulled off something, some kind of meld, some kind of blending.' He called it 'passion and pragmatism.'

And in July 2025, they ran RAADfest in partnership.[8] Strole and de Grey split stage time fifty-fifty, and de Grey programmed the speakers.

So if People Unlimited was such a 'disaster', suspect enough to be added to the list of organisations of concern by Rick Ross's Cult Education Institute, why have so many bought in? The easiest answer is that humans are terrible at minding red flags. The group managed to incorporate what could have been interpreted as failures – the death of its founder, for example – into the doctrine: the people who died did so because they didn't truly believe.

Death is always a threat to immortalists, writes Jeremy Cohen, a religious studies scholar at McMaster University in Canada who's studied RAADfest, People Unlimited, and the techno-immortal beliefs of Ray Kurzweil. If you die, you have 'not lived a sufficiently immortal lifestyle by eating well, taking the right supplements, being an active immortalist, and always cultivating a positive mindset.'[9] Dying is 'lazy', said Bernadeane on stage in the early 1990s;[10] it is 'not intelligent.'[11] According to Cohen, Charles's and other members' deaths are designated as 'suicide' because they were hiding death within, by not taking care of themselves, by being greedy, or being immoral in some other way. They weren't holy enough to live. The implication therefore is that leaving the flock will send you to your end.

Eternal life doesn't sound very relaxing; in the immortal universe, you might live forever with immortal anxiety. As an inverse of the sink or swim test in early modern witch trials, the only way to ever know if you are pure enough is if you don't kick the bucket. In the meantime, don't stop believing.

∞

Strole is now speaking with Naveen Jain. Jain is a serial entrepreneur who is impossibly absolute about the future of life extension.

'When you go up to them and say we're going to live forever, people say you're crazy. Great. Awesome. Right. Now, the thing is, what is really amazing about crazy people is it is easier to do the audacious things than to do something incremental, right?'

Jain punctuates his provocative statements with a rhetorical question at the end of every beat. The groupthink in the room makes it impossible to question the absurdity of what he's saying.

'People want to change something by 10 per cent. Well, we live to be seventy-eight. Maybe we could live up to be eighty-five, right? That's incremental. When you do something ten times better, that is when the radical stuff happens. We don't want to live to be eighty to go up to 120. We want to go from eighty to 800!'

The crowd throws their hands in the air. 'When someone else doesn't understand you, it is not your fault. It is their fault,' Jain reassures them. You can see the nods, the relief. 'They can't seem to go farther than their belief system,' he explains patiently. 'It is their mindset that limits them to the possibilities of what they can do.' Here he pauses for effect. 'Not what you can do. So I'm promising you immortality.' He looks around and catches the eye of someone a few rows back. 'What do you think?'

The audience sways.

'There is nothing written in our body that says you must die. Right?' Jain leans in. 'There is no reason and rhythm why we must die. Because at the end of the day, there is no mystery about this human body. It is simply biochemical reactions. And what if you could understand them? And if you can understand them, it's simply about math and chemistry. It's AI and chemistry. It is easy to fix, right?'

It's the scientists who are fools, he implies.

'I have a basic science degree from high school, right?' Jain says. 'When I graduated from high school, I learned that DNA makes RNA. Hey! And what happens in your body is determined by RNA gene expression that you can control based on what you eat, based on what you do, based on your lifestyle, based on your mindset – it changes everything.'

He pauses for the gratitude, and the applause.

Jain returns to the mic: 'Thank you. This is why I want us to go deeper with this mindset . . . For me, I want to go all the way to immortality.'

∞

Fast Company once described RAADfest as a 'semi-science fair for medical professionals who fall outside the norm of mainstream medicine'.[12] According to de Grey it's also building an audience of Silicon Valley VCs.[13]

But the geroscientists – whom de Grey describes to me as 'slow' – who are looking to attract funding from more traditional sources, are hesitant to attend an event that prioritises access to treatments over the research that underpins them. One exception is Irina Conboy. She and de Grey are colleagues; he was the first editor-in-chief of *Rejuvenation Research*, and she took over the role when de Grey stepped down in 2022. That was the same year she decided to enter the lion's den at RAADfest to set the record straight.

On the day she spoke, Irina was sitting at her desk in Berkeley, about 500 miles north. She was in the foggy Bay Area, engulfed by an oversized black turtleneck. It was Day 2 of RAADfest, and she was ready to address an audience of true believers. Some of them were Ambrosia Health veterans. She knew her science had led many to RAADfest, in the hope that she would provide the recipe for the philosopher's stone. She couldn't, of course.

Irina's headphones jumped to life. 'Let's make it happen!' shouted the voice in San Diego. 'Super longevity, unlimited lifespans! End of ageing and death! Everyone, long life, healthy life. Who wants a long and healthy life?'

Thunderous applause was amplified in her ears.

'... without the Conboys, we wouldn't have this process, and that's saying a lot!'

The video feed clicked on. She swallowed, adjusted her turtleneck and smoothed down her bangs as her CV was read out to the crowd.

'Professor Conboy has received numerous awards for her work in the ageing field, including Silicon Valley Community

Foundation, Open Philanthropy Award, Packer Endowment, et cetera, et cetera. Um, many, many awards. Um, very accomplished,' said the voice in her headphones. Irina raised her eyebrows as the presenter lost interest in her academic credentials. 'Let's see what Dr Irina Conboy has to tell us!'

The first slide was broadcast to the crowd, the title of her presentation: 'Old Plasma Dilution Reduces Human Biological Age: A Clinical Study.' Her face next to it was enormous and, unfortunately, her intergalactic virtual background wasn't able to cope with the clearance between her blonde hair and the headphones; there was a gap of beige reality poking through the trippy purple and blue hues. It was very distracting.

She moved to a slide of her data: several images with blue dots on a black background and, beneath them, several charts. These were the most recent biomarker measurements from the research in her lab that tests the rejuvenating effects of young blood plasma in old mice.

'ABT263 does reduce brain senescence, but does not improve neurogenesis and does not improve neuro-inflammation,' she explained, describing the results from a trial using Therapeutic Plasma Exchange (TPE – also known as plasmapheresis) to see if it would reduce 'senescent', or aged, cells. TPE is a process in which a person's plasma is exchanged for another substance – albumin, plasma modified with a treatment, or donor plasma. In this case, the patient's plasma was treated with a chemotherapy drug and returned to their body. 'But dilution of old plasma eliminates the senescent cells in the brain, and also improves neurogenesis and neuroinflammation – suggesting that dilution of old plasma is not simply dilution of inflammatory proteins or Sus proteins.'

Another slide.

'There's lots of data, but it's colour-coded. So the old serum is orange, and you see that either for P16, P21, IL6, matrix met-alloproteinase or diminished lamin B, all young cells become rapidly senescent after six days and cultured with old serum. And the same is true if you look at, say, beta gal or proliferation assays.'

A few coughs and creaks. Irina was telling a different, more nuanced story than the other speakers. She continued, unable to read the crowd from hundreds of miles away.

'Looking at the multidimensional UMAP projections,' she explained, 'here we have old people, control old people, and in red are participants who are before plasmapheresis. They are old, and you see they group closely with old people. But after the last round of TPE, they are far away from old, and they are close to middle age.'

For each participant, she explained, plasmapheresis resulted in a decline in the measured biological age. Plasmapheresis for age reversal in humans was Irina Conboy's science-fair entry. But it isn't the fountain of youth, she told me later. It's the fountain of middle age.

'Is it healthy or unhealthy? We don't know,' she added, trying to cultivate a note of caution. 'It is just one particular viewpoint on biological age, not the final determination of how young or old you are.'

The audience applauded politely as the rock music swelled. Irina was joined by another speaker on the stage: her colleague Dr Dobri Kiprov, an expert in TPE. The audience had a few questions.

'These are extraordinary people at the top of their game doing superlative work,' Strole reinforced to the audience. 'This is what the immortality lineage dictates: extraordinary, and great, results.'

He jumped in with a question: 'How often do you need to get a plasma exchange to maintain the benefits?'

Kiprov took the floor.

'It's a common question and very valid. We don't know exactly. My opinion is that everybody is different, and people will need different schedules. At the moment we're evaluating three different schedules with the clinical trial that I discussed. So hopefully we'll have some answers by the end of next year.'

A question came in from the audience.

'Would just donating blood regularly result in the same improvement?'

This referred to a rumour going around online that taking a trip to the blood bank worked just as well as plasmapheresis, because when you donate your blood, plasma is removed at the same time as the red blood cells. The rumour claimed that this 'cleansed' the body of the substances in plasma that cause ageing, at a fraction of the cost of plasmapheresis.

'Very common question,' Conboy responded. 'In fact, I received numerous emails.' She slowed down to make the next statement very clear. 'I don't think that donating blood will work because you need to donate so much blood that you would die from catastrophic blood loss in order to dilute sufficiently.'

She continued.

'There is no shortcut . . . I do not recommend biohacking or donating blood or plasma.'

There was one more question, this time from an entrepreneur at the back.

'What are the regulatory requirements for doing apheresis, plasmapheresis, in an office setting?'

'I would not recommend that,' said Kiprov. 'Ninety-nine per cent of the procedures are done for really ill, sick people with life-threatening conditions. Most of those are done in an ICU setting. Some require plasma, which can cause all kinds of problems and medications. So, moving this from the ICU into the office is not advisable. I think it is extremely unsafe. It does not do anybody any good. So if your corner med spa suddenly hangs a shingle that says, get your plasmapheresis here, run very fast. Very fast.'

Strole laughed. 'Anyway, I want to thank you guys. Thank you so much. I can't wait to get in there and get some plasmapheresis.'

Strole had taken control of the narrative. He left the stage in the direction of the Expo Hall, and Irina Conboy's Zoom cut to black.

∞

In April 2023, a few months after Kiprov had advised the RAADfest audience against getting plasmapheresis from a med spa, Bryan Johnson dropped into a longevity clinic to test whether young plasma was worth adding to his Blueprint Protocol.

'We have pretty rigorous biostatistical criteria,' he tells me, referring to his thirty-strong personal medical team. 'We have an orientation on what therapies offer, what potential benefits, what potential side-effect profile, whether it's an animal model, a human model, like, you know, all the variables you'd want to consider.

'We thought plasma was interesting, not because of the overwhelming, compelling evidence, but because of the potential benefits,' he explains. 'There were quite a few people who were enthusiastic about the practice, and we heard a lot of anecdotal reports of subjective feelings of wellness.' In a case of life imitating art, he and his team found a 'blood boy' that fit their criteria, preferring to oversee his diet, exercise and lifestyle rather than selecting from a blinded supply chain. 'You didn't know if people were obese, you didn't know if they had high inflammation,' he tells me. For his purposes, he needed a pure, grade A matched donation so he could run the numbers on any changes to his biomarkers. They performed TPE with the donor's plasma in place of Johnson's for five months. After that five-month series, he decided to get one more month's worth of plasma, but not from his usual donor. Instead, he sought blood from his teenage son, Talmage.

'I, of course, have the best intentions for my son for his health and wellness,' Johnson says to me, 'but I've never paid as close attention to what he's eating as I did prior to this plasma exchange, because that's going into my body.' He laughs. 'He was a proxy for my own existence.' Talmage's biological processes were going to be Johnson's. 'It's a whole different level.'

He takes me through the process.

'When we arrived at the clinic . . . I think there was a lot of, uh, anticipation. I mean, Talmage had never had this done before.'

Johnson describes the scene. In the clinic, he took a seat. 'Talmage has had his blood drawn,' he says, 'but it's entirely different when you're hooked up to this machine and it's drawing a lot of blood out and it's spitting out the plasma.' The 17-year-old sat next to his father in the other reclining chair. 'It's a very involved procedure from just a small poke. He had a lot of questions.'

The machine started to whir and Talmage's arm was wiped down with iodine. The needle went in, and his blood made its way along the tube into the centrifuge, where it spun at high speed until the red blood cells separated out, leaving Talmage's healthy, transparent, yellowish plasma. Another tube ran the red blood cells back into Talmage's other arm. The plasma was deposited into an IV bag. Talmage sat, sanguine, for the forty-five minutes it took to extract a litre.

Then it was Johnson's turn. It was obvious he was enjoying telling me all this. 'It's like you have your innards on display, and people are going to rank and assess them.'

So how did he do? His medical team and the team in the clinic crowded around his and Talmage's full bags, comparing them and murmuring, 'What do you think? Does it look healthy?'

Yes, it did. Johnson's 45-year-old blood plasma was as clear as his teenage son's. Johnson lay back on the chair while Talmage's plasma was injected into his body.

Meanwhile, Johnson's father, Richard, took Talmage's seat and was prepped to receive *his* son's plasma. Johnson had re-forged a relationship with Richard after he'd left the church, and Bryan had been sharing his Blueprint Protocol with his father. 'My dad called me one day and he said, "Hey, Bryan, I just want you to know that when you begin experiencing cognitive decline, you don't know it. I always imagined I would feel myself slipping and say, I'm just not as sharp as I used to be." You don't know, which makes sense: it's decline,' Johnson comments. 'Therefore you have this blind spot to your own decline.'

This was the real experiment: was Johnson's blood young enough to reverse his dad's decline? As the valve was opened and the infusion began, his dad started crying. 'This was one of the most meaningful moments of his entire life,' Johnson tells me.

'And for me too. I've done these plasma exchanges before, but never from my son. I never imagined that my little baby would grow up and I'd be doing this procedure with him.'

A few months after he relayed this story to me, Johnson tweeted an update to his 484,000 followers:

Discontinuing therapy: completed 6, 1L young plasma exchanges. 1x/mo (1 w/ my son). Evaluated biomarkers from biofluids, devices and imaging, no benefits detected.

Young plasma exchange may be beneficial for biologically older populations or certain conditions. Does not in my case stack benefit on top of my existing interventions.

Alternative methods of plasma exchange or young plasma fractions hold promise.

My father's results still pending.[14]

In November 2023, Johnson tweeted again, claiming that one litre of his 'super plasma' might have been the thing that slashed Richard's rate of ageing by twenty-five years. 'The older we get, the faster we age. After receiving 1 L of my plasma, my father is now aging at the rate of a 46 year old. Previously, he was aging at the rate of a 71 year old. I am my dad's blood boy.'[15] The FDA's 2019 caution against using young plasma as a treatment for age still stands; by June 2025, there had been no further updates on Richard's health from Johnson.

In October 2024, Johnson tweeted that he was starting TPE, replacing his plasma with protein-rich albumin. 'The therapy objectives are to remove toxins from my body. The evidence is emergent,' he wrote.[16]

How should we face death? Denying it, like Strole and Jain? Or fighting against it, like Bryan Johnson and the geroscientsts? Or accepting it? Lucretius, who is credited with bringing Epicurean philosophy to Rome, wrote in the middle of the first century BCE that 'Death is nothing to us. We shall not be conscious after death any more than we were before birth. Death, then, is naught to us, nor does it concern us a whit, inasmuch as the nature of the mind is but a mortal possession.'[17] This would be a compelling argument, except for one important thing: we can see death happen to other people.

Stephen Cave describes this as the essence of our 'mortality paradox':[18] we can't imagine a time when we are no longer here, but we know that it will come because we have experienced the loss of others. So in our grief and existential meltdown, some choose to reject death and pursue immortality.

Charles Paul Brown first experienced a cellular awakening after his first wife died in a car accident. Aubrey de Grey inherited $16 million when his mother passed away, $13 million of which he has dedicated to ageing research.

In 2024, Bernadeane died of breast cancer, just before the ninth RAADfest. It was at this gathering that biohacker Dave Asprey presented his newly released product on the main stage: the Wasabi Method, a shock-wave therapy for tissue repair and rejuvenation. That year, there were speakers who declared the benefits of TPE, and Strole introduced two new full-length conference streams: 'Longevity Activism', to galvanise the fight for the rights of patients to choose what treatments they should be able to take, and a 'Women's Longevity Lab' that combined 'spirituality and science' so women could 'shift their health and shift their consciousness'.[19]

Bernadeane continued to deny her end, even after she died; her body was cryogenically preserved. Throughout her life, she kept faithful to the immortality message, but it evolved: from Christianity into a credo of the immortal lineage, ultimately becoming a conviction in science and technology. At each of these junctures, she believed death was in her control. And so do the faithful immortalists of Silicon Valley: they are convinced, beyond a shadow of a doubt, that all they need to do to solve our mortality is to set their machines on the problem and let *them* figure it out.

This is their religion.

PART III

Post-Mortal

CHAPTER 7

The Techno-fundamentalists

Sci-Fi Author: In my book I invented the Torment Nexus as a cautionary tale.
Tech Company: At long last, we have created the Torment Nexus from
classic sci-fi novel Don't Create the Torment Nexus.

Alex Blechman, writer at *The Onion*[1]

'When was the moment you realised you didn't want to go for
pizza and a beer after work?' I ask Bryan Johnson, who's sitting
serenely in the half-darkness of his giant living room. I know
when he stopped, but what I want to know is when he got over
the cravings. Johnson quietly smiles. I sense a capital-t Truth is
about to be told.

'I've become the most measured person in history,' he says
solemnly. 'The data is outputted and referenced to scientific data
and a protocol is created, and I follow that protocol with perfection. My mind is not involved. My mind cannot look at a menu.
It cannot peruse the pantry. It can't participate in a spontaneous
pizza party.' He pauses. 'I've built my autonomous self.'

Building an autonomous self is not the quick fix I'm looking
for. Nor is devolving responsibility for my wellbeing to an expensive algorithm. I admire Johnson's willpower; he lives around
the corner from fancy restaurants, and is 15 minutes' walk away
from a world-class farmer's market where you can buy all kinds

of delicious things. But instead he chooses to only eat Nutty Pudding (made with various nuts, pomegranate juice and cocoa), Super Veggie (a blend of steamed vegetables, lentils, vinegar and spices), or 500 calories of some combination of veggies, nuts, seeds and berries, because that's what his computer tells him to do.

Now I like nuts and vegetables probably as much as I like pizza and beer, but if I was told I could only ever eat them again, I wouldn't be able to do it, no matter how much extra life it gave me. Every day, forever? No thanks. I don't have a fraction of his faith in technology. Because that's what it all comes down to in these immortalists' lives: each – through a combination of fantasy, wishful thinking and mathematical reason – believes that technology is the thing that will give them eternal life. Some – like Johnson – are using it as a tool to get there in a form that is very much human. Others believe we must and will all merge with it, to become 'post-mortal'. All are what I call techno-fundamentalists.

As we continue to become intertwined with our digital tools, we increasingly believe that technology is in control, for better and for worse. Those who think it acts in our best interest – like Johnson – prefer to let it take charge. 'I'm trying to demonstrate there may be a new step in the evolutionary path here where our minds are not the unchallenged authority,' he tells me. So, for Johnson, technology's first challenge is to solve the fact that every day he is getting one step closer to six feet under.

When I spoke with him in 2023, he was forty-five years old. His Blueprint algorithm already helped him 'reverse' time, by devising a protocol that would bring his biological age signifi-cantly lower than his chronological age. But now he wanted to see if he could stay the same biological age for one year.

On his forty-sixth birthday, he tested his biological age

again, using a biomarker tool called an epigenetic clock, and it appeared that his computer was indeed doing its job: his body was ageing at a rate of 277 days for every 365.

Now, holding back time for one year may not seem enough to live forever, but here's why it might: think of it like a rocket leaving the earth. It needs to reach a certain speed to overcome the pull of gravity. This is called escape velocity. Now imagine that through treatments, lifestyle changes and prayers you could stop yourself from ageing. The idea is that you would eventually reach the point where each age-related disease, such as dementia or diabetes, and those that we don't know about yet, will have been cured before your body ages into them. You would escape the pull of mortality. This is called longevity escape velocity.

Fundamentally, this isn't living forever; it's not dying today. De Grey coined the term in 2004, and now runs the LEV Foundation, a non-profit dedicated to researching and developing medical treatments to prevent and reverse age-related diseases. Over the years he has converted others to his cause, like Johnson. Entrepreneur and biohacker Peter Diamandis is also trying to stay the same age forever. Diamandis, writing on his blog in 2024, is sure that 'in the near future (at some point), additional scientific breakthroughs will extend your lifespan by more than a year for every year you remain alive.'[2] But not everyone is convinced.

'I understood certainly even before I started talking about thousand-year lifespans on stage that it would frighten the horses,' de Grey says to me in 2025. 'But what I did not expect was for all of my colleagues to a man to cut out of that and to just pretend that they couldn't do the utterly elementary mathematical reasoning that leads to the concept of longevity escape velocity once one has rejuvenated people to the extent of giving them even twenty years of extra life.'

The escape velocity cusp is closer than most of us would imagine, he maintains. In 2004, he predicted that, as we are already so long-lived, 'even a 30% increase in healthy life span will give the first beneficiaries of rejuvenation therapies another 20 years – an eternity in science – to benefit from second-generation therapies that would give another 30%, and so on ad infinitum.'[3]

'And even now,' he tells me, 'the people who have decided that they can't deny this any longer still pretend that it's obviously absurd just because the conclusion, it sounds absurd – irrespective of whether the reasoning is impregnable.' Forever is a process – and one which de Grey believes in.

Now, longevity escape velocity relies on a major mathematical assumption: that technology can keep up with the heterogeneity of human decline. What this means is that conditions that affect you when you're young are – medically speaking – relatively simple. But the illnesses of old age could be said – loosely – to grow in severity at an exponential rate. Johnson has to stay young, or an age-related disease that can't yet be treated will kill him.

If you are trying to age in reverse, you would need to be living in a time when treatments and technologies both improve at the same exponential rate. These people are literally racing against the clock and are really, truly trying to win. De Grey's reasoning for why they might is based on a 'wild extrapolation' made about technology in 1965.

Gordon Moore's name is on buildings and plazas of university and corporate campuses worldwide. Mention him in a roomful of computer scientists and futurists and their eyes go misty. His name is an incantation. It means, 'Everything is predictable.'

Moore was a physicist by training but an inventor,

entrepreneur and philanthropist in practice, working on computers' tiny but mighty brains. He was part of the team that picked up one of the very first VC investments in Silicon Valley – the group of inventors known as the 'traitorous eight' who had left the Shockley Semiconductor Lab en masse to form the influential company Fairchild Semiconductor in 1957. Ten years in, Moore and his colleague Robert Noyce jumped ship from that venture and, with the help of their VC friend Arthur Rock, co-founded the Intel Corporation in 1968.

It was as head of research and development at Fairchild that he was asked by the journal *Electronics* to contribute a now-famous article predicting the next ten years of his industry. On 19 April 1965, the journal published 'Cramming More Components onto Integrated Circuits', in which Moore imagined that 'home computers – or at least terminals connected to a central computer – automatic controls for automobiles, and personal portable communications equipment' would all be invented within the near future because of the economics of innovations in computer chips.[4] 'Semiconductors are an unusual technology,' he observed in an interview looking back at his career in 2014. 'By making things smaller, everything gets better at the same time: the transistors get faster, the reliability goes up, the cost goes down. It's a unique violation of Murphy's Law.'[5]

He plotted this out: every year, the number of components you could fit onto an integrated circuit (a computer chip) for a given price doubled. That's double everything that powers a machine: the sensors, computer memory, computer processors, amplifiers, timers and counters. Or, he argued, you could also see the inverse: the price for a certain number of components on a chip halves. That means that the number of components doubles every year.

It helps to break this down into numbers: in year 1, when

you have 1,000 components, your chip will cost you $100. In year 2, you'll get 2,000 components on your chip for $100. In year 4, 8,000 components for $100. And by year 8, it'll be 128,000 components for $100. If you want a chip with the original 1,000 components in year 8, it'll cost pennies.

To Moore's surprise, when a decade had passed, his prediction 'turned out to be ridiculously accurate.' The semiconductors that he was building at Fairchild and later at Intel doubled in complexity while doubling in efficiency. As they were added to washing machines and network servers, tablets and supercomputers, they came to underpin how the world works, connects and consumes. It's no understatement to say that Gordon Moore's great idea is another foundation stone of Silicon Valley.

By 1975, people had started talking about the number of components on a microchip as a proxy for computer capability – that things got 'twice as good.' They'd dubbed this phenomenon 'Moore's Law', much to Moore's embarrassment. He made a few tweaks to his timeline, adjusted the rate of exponential increase from two years to eighteen months, and today Moore's Law is the reason the phone that you have in your hand costs around the same as the last model (not taking into account inflation), but is twice as powerful.[6] An integrated circuit is roughly 2 billion times more powerful today than in 1960.[7]

For the record, it's not a law – it's an observation that continues to be constant. But Moore's calculation has been true for such a long time that it's now become technological *lore*. It drives every innovation and every flow chart. It is the basis of both dreams and strategy. Moore's Law sets the pace, so developers can imagine a predictable timetable of a leap in computing every eighteen months, so they can plan for their next leap in technology. Everyone is working to this timeline, so everyone is pushing everyone else's envelope. It is inconceivable that anyone would say Moore's Law will fail (though there is evidence it is slowing[8]), because everyone's business model is based on it. Verily, the emperor would have no clothes.

Moore's Law is now also being applied to modern medicine. Everything from MRI scans to drug discovery to protein folding is regarded as a computational problem, therefore it too should be characterised by an exponential increase of capability. If modern medicine is subject to Moore's Law, every eighteen months it gets twice as powerful at keeping you alive, so what will medicine be like in twenty years' time? And if you figure out how not to age in that time, you begin to understand why living to 200 years seems like something that could happen.

This is not science fiction, though for many years writers have been predicting that the moment when we successfully outrun time will arrive in the near (to their) future – regardless of when they were writing. Nonetheless, sci-fi has always been a hugely influential genre among geeks, and it has inspired all kinds of inventions, from the first smartwatch, released in 2015 (based on an obsession Apple CEO Tim Cook had with the comic sleuth Dick Tracy), to the first commercially available cell phone, Motorola's DynaTAC (inspired by *Star Trek*).

You might think I'm placing too much emphasis on a genre of fiction, but the stories from our childhoods determine what we envision as desirable and possible. Jonathon Keats is an experimental philosopher, and the founder and curator of the Museum of Future History. 'The Museum of Future History is an institution devoted to exploring how past visions of the future have influenced today,' he told me for *The Digital Human* in 2021. The museum's first exhibition was *Toying With Tomorrow: Playthings That Anticipated the Here and Now*, bringing together toys from all over the world. It was timed to coincide with UNESCO's High-Level Futures Literacy summit. 'When we look at toys, we are looking at what children envision in the future,' Keats explains. He cited a growing concern among futurists that the visions of the future that we have 'colonise' that future – they predict what it will bring, potentially limiting the possibilities of future generations.

'Each generation really needs to be able to make its own decisions,' he says. 'We don't know what sort of world the gener- ation after us will live in. There's a moral imperative that we not occupy the future in a way that limits the autonomy of the next generation. And we make it a lot harder when we give children a vision of the future that is so suggestive that they take it as their own.'

The toys exhibition was a way to make those ideas less abstract, looking at the whole range of playthings in a given time, 'and how that sets certain patterns of expectation.'

I came to realise just how much this is true when studying an early version of a 'metaverse', or virtual world. It was – still is – called *Second Life*, a noughties graphical sandbox that looked like a computer game where people who used it 'wore' avatars that moved around the digital landscape and built the environ- ment around them. Think *The Sims* meets *Roblox*. It was massive and multiplayer – and anyone could come to what you built and visit, hang out, talk and co-create. It was like a chat room, but with pictures.

This was 2003, and the company who produced *Second Life* was called Linden Lab. It was fronted by a developer I got to know named Philip Rosedale. He recommended a book by the sci-fi author Neal Stephenson called *Snow Crash*, which described essentially what *Second Life* was: an immersive virtual environment where everyone on screen built the world and themselves.

This was just before the blockbuster *World of Warcraft* was launched, the era of *EverQuest* and *Asheron's Call* – influential titles that have helped to define modern online role-playing games. What struck me was that, although anything should have been possible, everyone just imported online the social systems they already knew – economies and beliefs about what was valuable or not, ideas about civil society and justice, and hierarchies of power. Even in a totally free environment, where the possibilities should be endless, we import our past experiences to organise ourselves, and this is what inspired my PhD in social psychology.

'Science fiction influences everything in this day and age, from the design of everyday artifacts to how we – including the current crop of 50-something Silicon Valley billionaires – work,' explained Charles Stross in 2023.[9] If they imagine a future in which we will live forever, they will try and build it. Stross likes to write fiction about what he imagines are the utopias of inventors. For example, at the centre of his award-winning 2005 book *Accelerando* is a communication network of superintelligences in a post-scarcity society in which 'the logic of competition pushes [capitalism] so far that merely human entities can no longer compete.'[10] He wrote it while working as a programmer, and as the dot-com bubble was inflating. The company he was working for was growing at a remarkable rate of 30 per cent per month, and there, exponential growth meant, 'the workload is always growing faster than the budget for hiring minions to do the donkey work,' he wrote on his personal blog in 2010. Stross stepped away, and imagined what would happen if the world continuously accelerated at that rate. 'What kind of person can [keep pace] while all around them the world is melted down and re-forged monthly, daily, hourly?' he asked himself.[11]

This is the Law of Accelerating Returns, an offshoot of Moore's Law. It was dreamed up by Ray Kurzweil, the RAADfest speaker who imagined we would become immortal through merging with AI. He believes that whenever technology bumps up against a barrier, another technology will be invented to overcome it, and with every generation's improvements, the downstream effect is one of accelerating returns. Kurzweil imagines that this will lead to 'technological change so rapid and so profound that it represents a rupture in the fabric of human history.'[12] This is why, according to Kurzweil and his techno-fundamentalist disciples, ever-advancing technology will be able to solve the ever more complex problems we throw at it, including humanity.

But there's one very important thing to remember: technology is not agnostic. It directs us through the digital world. Sometimes it does so clearly, with buttons or features. Sometimes, it's not so obvious. For example, my personal mental algorithm around food is, 'If hungry, then eat.' Johnson's, for the same situation, is, 'If hungry, then refer to the Protocol to see if the algorithm says it's time to eat. If yes, then refer to the Protocol to see what the algorithm says I can eat. If I have those ingredients, then eat.' Problem: hunger. Win condition: get food. Algorithm: how it gets to the win condition.

Two different technical algorithms built to solve the same problem will vary. My version of a machine that answers the question 'how do I live forever?' could be different from someone else's, and this would prioritise some things and render others invisible. Algorithms – including the technical ones – are imbued with the intentions of their creators, which defines the pathway they follow to find a solution.

Charles Stross believes tech developers are motivated by 'the American dream of capitalist success, combined with uncritical technological solutionism and a side order of frontier colonialism' because this was the content of the sci-fi popular from the 1950s to the 1980s.[13] You can see this in how technologies have been rolled out over the last three decades: Uber launched operations in cities without getting prior approval; Google's Street View started taking photos without considering privacy concerns; Facebook collected data without consent. They do this for market penetration – breaking things first, and asking for forgiveness later. The people wielding the power, in the era of what Greece's former finance minister Yanis Varoufakis described as technofeudalism, are the 'tech plutocracy' – the owners of the companies that he accuses of warping the economy and setting us against one another.[14] The thing is, if tech continues to develop apace, as per Moore's Law, and as its proponents believe, rapidly accelerates upon its returns, as they also believe, we're heading towards a sci-fi outcome that they *can't* control.

In the 1980s, Vernor Vinge, a mathematics and computer science professor at San Diego State University and sci-fi author, posed a question in an article in science-fiction magazine *Omni*: what if the exponential growth of technology never stops? This was a popular cultural trope at the time. He imagined a robot that could think – an artificial intelligence becoming rampantly smarter with every passing year. If – always if – an artificial intelligence was even possible, thanks to Moore's Law we would be able to make it run faster with ever more advanced computing power, and problems that were once perceived to be impossible would be quickly solved. 'The evolution of human intelligence took millions of years,' Vinge wrote in 1983. 'We will devise an equivalent advance in a fraction of that time. We will soon create intelligences greater than our own.'[15] And when this happens, it will continue apace, and irreversibly. We won't be able to second-guess it. It will always know more. We will have no choice but to merge with it, forming a singular post-human, artificially intelligent entity. And the moment when this happens will be known as the Singularity.

Given that humans are theoretically dumb, and appear to be only a little more advanced than other species, what would happen, Vinge postulated, if you took this smarter-than-human artificial intelligence and challenged it to solve 'us'? Immortality of a fashion, of course.

'But there is a problem,' explained Stross. 'SF authors such as myself are popular entertainers who work to amuse an audience that is trained on what to expect by previous generations of science-fiction authors. We are not trying to accurately predict possible futures but to earn a living: any foresight is strictly coincidental.'

Techno-fundamentalists believe the story though.

'The billionaires behind the steering wheel have mistaken cautionary tales and entertainments for a road map, and we're trapped in the passenger seat,' Stross warned.[16]

To the people who have faith in Moore's Law, the Singularity is in sight. And with it *will* come eternal life.

CHAPTER 8

Nerd Rapture

Ten thousand years ago, the state-of-the-art was a goat.   ✗

Cory Doctorow, *Down and Out in the Magic Kingdom*

The end of the world should have happened at midnight on 1 January 2000. That was when a computer bug, Y2K, was predicted to cause a Silicon Valley-engineered apocalypse. It was going to topple electrical systems, transport systems, financial systems – anything that had a computer chip. The moment we ticked over to the new millennium would be a chaos of our own creation.

The coders who'd created the early mainframe systems in the 1950s and 1960s hadn't anticipated the future. They were operating with expensive memory, so they optimised their machines to record dates as two digits: 1962 became 62; 1991 became 91. They didn't imagine that their computers would be around as long as 00. But as we got closer to the year 2000, and our civil societies had intertwined with computers, the theory was that if the systems that relied on these dates were unable to discern between 2000 and 1000 or 1600, they'd throw up an error and operations would stop until that error was fixed. Unfortunately, some of the systems were very important: doors at high-security prisons,[1] the automated movement of global currency, airplanes.[2]

Everything, the doomers predicted, was going to shut down. After a *Y2K Family Survival Guide* – narrated by Leonard Nimoy, *Star Trek*'s calm and collected Mr Spock – was released on VHS, people panic-bought food, water and toilet paper.[3]

In the years leading up to the first bells of the new millennium, wizards around the globe furiously re-engineered the code to stave off this hypothetical total societal meltdown. Research firm Gartner estimated that it would cost between $300–$600 billion to fix the 'millennium bug'.[4] The US government passed the Year 2000 Information and Readiness Disclosure Act to encourage companies to share their progress. The United Nations convened two Y2K conferences to coordinate an international response to the crisis.[5] An army of coders picked up consultancy work for remarkable fees.

Many in Silicon Valley would say that we are here in the second decade of the twenty-first century in part because of their efforts. Perhaps nothing major happened because the response was so great, or perhaps the threat was massively overblown. The enduring moral of this story is that only Silicon Valley could save the day, and for most people this was the first time they got a taste of how much the world had changed, and of how little the people who had until this point held power really knew about the computerised world.

When powerful technologists like Ray Kurzweil, Elon Musk, Marc Andreessen, Sam Altman and others talk about the inevitability of the Singularity, or how AI is an uncontrollable force, it feels like we are *actually* on the cusp of a Silicon Valley-engineered apocalypse propelled once again by the inability to predict the future – and it's the moment the techno-fundamentalists have been waiting for: 'We are baby steps away from creating superintelligence, potentially the most extraordinary event in the entire galaxy, and it happens to be in the very moment that you and I are alive,' according to Bryan Johnson. 'I would not miss admissions to this show for anything.'

Why talk about AI in a book about immortalism? Because apparently this tool is the one that's going to actually, finally, decode humanity, thus rendering both our minds and our bodies solvable. 'When I look at the speed of technological progress, I don't know how any human could arrive in a reasonable fashion at a cap of lifespan,' Johnson tells me.

The difference between Y2K and AI is that the former was a computer glitch that some believed had the potential to destroy humanity, and the latter is one that its creators, based on Kurzweil's Law of Accelerating Returns, believe will give us immortality (if it doesn't destroy humanity first). For them, it's a logical next step.

Imagine what it must be like to be an old hand working in technology today. Maybe you're wealthy, maybe you're not. Maybe you're an engineer so deeply entrenched in the edge-cases and fine details of your products that you find speaking to others a jarring experience, or maybe you're a visionary, full of evangelical zeal, if short on technical knowledge. Either way, the past decades have been quite a ride.

The journey to here will have left you with a few impressions. First, that technological capability will continue to improve, as it has done for hundreds of years. Second, that the rate of improvement is exponential, as it has been since at least the 1960s. Third, that you are one of the few people who knows what it all means. Or at least, you know what it *meant*.

Moore's Law used to mean a faster machine, or maybe some better graphics, or a niftier console: as with crop yields, or weaving, we would use words like faster, stronger, more productive. Predicting the future used to mean: the same, but faster and cheaper.

But now, in the 2020s, Moore's model of predictability has

broken. The accelerating returns found in processor design, data storage, memory efficiency and transmission speed, combined with the equally accelerated returns of online publishing and cultural creation – the way we've put all of our knowledge online for the past twenty years – has given us AI.

AI has been around for a long time, but its rate of acceleration in 2025 is far faster than it was before: our ability to get computers to at least appear to think, to reason, to solve problems, seems today to double every few months, if not weeks.

So what happens if someone who was raised in Gordon Moore's Silicon Valley witnesses these new tools through the lens of their experience of the past quarter century? They extrapolate. If today's large language models are the AI equivalent of Atari's 1972 computer game *Pong*, what does the AI equivalent of *Call of Duty* look like?

Developers such as Altman, Musk and Google's Nobel Prize-winning DeepMind AI department think they know. Or, at the very least, they think they know what it will *be*, but not what *form* it will take. What's standing in the way of their vision of the future is less about the technological capabilities – they will invariably accelerate. No, what's blocking them from the next great leap forward is the puzzle of mortality that Leonardo da Vinci and Descartes wrestled with centuries ago, which today's developers are, of course, trying to reverse-engineer.

AI has for a long time outperformed humans in certain cognitive capabilities that are what we, culturally, have assumed to be a measure of intelligence. We can go back to 1966, when Joseph Weizenbaum invented a 'chatterbot' which he named ELIZA. ELIZA was one of the first natural language processing programmes: it answered questions that a user posed in everyday language. It mimicked a conversation on the surface but, under the hood, the programme was using pattern matching to solve the problem it had been set: it applied rules created by Weizenbaum to determine what in its archive of scripted responses would make a user think it was having a conversation with a machine. He modelled it on the psychotherapeutic techniques of Carl Rogers: listening for keywords and then asking for clarification in the form of questions using those keywords. When I interviewed Weizenbaum's daughter in 2023 for an episode of *The Digital Human*, she told me that some people were so taken in that they actually mistook the bot for human.

Fast-forward to 1997, and a supercomputer called Deep Blue beat grandmaster Garry Kasparov at chess. It also used pattern matching, and applied technical brute force to determine each possible move. In wildly simplified terms, it played an entire game of chess with every turn.

Only twenty years later, in 2016, Alphabet's AlphaGo beat Lee Sedol, one of the world's top Go champions, in four out of five games by 'learning' how to play. The bot had been trained by playing against both humans and itself; unlike Deep Blue, it didn't brute-force it. It moved *strategically* based on reinforcement learning and an artificial neural network to 'learn' from its own mistakes and adjust accordingly.

It wasn't the win that surprised the chess and AI research communities. It was that AI played a totally inexplicable move. It didn't invent it from scratch, but it was one that only had a 1 in 10,000 chance of being played. It was a creative solution. It was a beautiful move. It became legendary.* *Wired* described it as 'a moment of genius',[6] and Lee subsequently described AlphaGo as 'an entity that cannot be defeated'.[7]

In 2024, the Nobel Prize in Chemistry was won by Demis

---

* Though, it should be noted that in the fourth game Sedol also used a move that had a 1 in 10,000 chance of being played, which also shocked the crowd, and earned him his only winning round.

Hassabis and John Jumper, part of the same team at Google DeepMind. They had developed an AI called AlphaFold that 'solved' a fifty-year-old unsolvable 'problem' in biochemistry. AlphaFold was an even more advanced machine than AlphaGo because it used a learning technique called generative AI. Generative AI takes massive datasets and trains itself with very few – if any – parameters set by humans. In other words, it internally figures out its own connections and then it produces new content that is based on what *it* learned about the existing data. The team created a model that predicted the 3D shapes of 200 million proteins, greatly outperforming other measures.[8] This outcome has rocked biology and medicine; it gives researchers a proverbial decoder ring to understand proteins' functions, design new drugs, and develop new materials.

Generative AI is the same learning technique that powers the current generation of large language models, like ChatGPT and Claude, as well as AI image and video generators (like Midjourney, Sora and Stable Diffusion) and music generators (like Suno AI), and though it is extraordinary, it is what Kurzweil describes as 'narrow AI': amazing at doing the one thing it's trained for. It can become a clever chatbot pretending to be a human, or a virtual voice assistant, or even a self-driving car. But it can't become a self-driving car that is *also* a chatbot *and* a protein-folding machine *and* a Go grandmaster. To be that, an intelligence would have to be even more like a human because it would have to have the capacity to learn a lot of things. And apply them when it decided it needed to. It would have to be *generally* intelligent. And this is the next goal of Silicon Valley: Artificial General Intelligence, or AGI.

Creating an AGI is what Altman and Musk set out to do when they and a consortium of researchers, scientists and entrepreneurs co-founded OpenAI in 2015, funded by Peter Thiel and others, and what Microsoft invested in when it put $10 billion in the company in 2023.

'The launch of ChatGPT kicked off a growth curve like nothing we have ever seen – in our company, our industry, and the world broadly,' wrote Altman in 2025.[9] But what the company is aiming for is an AGI system that would have some kind of internal self-awareness and self-understanding. This would mean that once it's up and running, *it* would determine what needs to be learned and then *it* would decide if it's a good idea to go and do it. The difference is that AI *simulates human learning*, while the hypothetical AGI learns *like a human*.

Enter Engineer's Syndrome: there are infinite ways to decode human intelligence, and numerous debates about what learning is and how it happens. But engineers in the Valley are trying to decode the electrical pulses of human thinking at the level of the neocortex so they can then reconstruct them in a machine. Others are focusing on a 'mathematical essence' of general intelligence, and plan to apply that.[10] Still others are trying to pin down the processes involved in symbolic thought. This techno-solutionism is drawing investment from VCs: globally the market value of the AGI development industry was $3 billion in 2023; 39 per cent of that was in North America. This figure is projected to increase to $52 billion by 2032.[11]

'Technology brought us from the Stone Age to the Agricultural Age and then to the Industrial Age,' wrote Altman on his blog in 2024. 'From here, the path to the Intelligence Age is paved with compute, energy, and human will.'[12] The visionaries' promise is that their will, and their technology, will take us to AGI.

Once we have AGI, the prophets predict we should be able to discover all kinds of new treatments for ageing. Altman believes we are close. 'We can now imagine a world where we cure all diseases, have much more time to enjoy with our families, and can fully realize our creative potential,' he wrote in 2025.[13] So as a first step, yes, this is what could buy us time until we can achieve longevity escape velocity. This is also how a fringe philosophy once imagined on the Extropian mailing list has now captivated many of the leading lights of Silicon Valley. Transhumanism has arrived.

Nick Bostrom, whose parable of the Dragon-Tyrant we came across in the Introduction, is the father of the modern rebirth of the transhumanism movement, which came into its own in the 1990s and early 2000s. 'It's the idea that we can and should use technology to enhance and expand and augment human capacities, whether it's extending the healthy human lifespan or cognitive enhancements, ways to get more control over emotional capacities, or enhancing the human body,' he explained to me over the phone in 2023 from Switzerland, where he was writing his latest book, *Deep Utopia: Life and Meaning in a Solved World*. For more than thirty years, he has been investigating what might happen with the advent of potentially transformative technologies. 'You could say it is almost an obsession,' he tells me.

According to Bostrom, the path towards radical life extension won't go through biomedical research. It will instead go through AI.

I ask him if he wants to live forever.

'I think I'd want to have the ability to postpone making that decision,' he says, characteristically. 'I think you'd need at least a few thousand years to ponder this question before you would be ready. But even then, I think it would be better to make it piecemeal, this choice. You decide, "Would I prefer to die now or live another year?" And then if you choose to live another year, you can revisit that question, and if the end result is that you live for a very long time, then so be it.'

Bostrom co-founded the World Transhumanist Association (now Humanity+) in 1998, to create a counterweight to a prominent idea emerging in US politics in response to innovations in genetic technologies that we should not augment our natural abilities. Transhumanists believe in morphological freedom, that a person should have the autonomy to change their body and their mind in whichever way they feel like, and they should pursue technology that will help people find ways to achieve fulfilment. This might include people who implant computer chips and sensors under the skin, or, Bostrom says, it could include 'a bunch of folk who would look quite eagerly on the prospect not just of adding ten more healthy years to their human lifespan, but to radically transforming themselves by uploading themselves into a computer and then increasing the size of their brain to a planet-size super brain, and then living for billions of years.'

In 2005, with funding from computer technologist and author James Martin to address problems facing the future of humankind, Bostrom established the Future of Humanity Institute (FHI) at the University of Oxford, which, until it closed in 2024 under a cloud of accusations of being 'cultish' and 'toxic', drew substantial funding from Elon Musk, the Future of Life Institute (FLI) and Facebook co-founder Dustin Moskovitz's Open Philanthropy fund.[14] While he was there, Bostrom formalised the canon of transhumanism, writing extensively about the social and political power of technology to be a driver towards utopia.

Meanwhile, Kurzweil was appointed Google's Principal Researcher and AI Visionary and one of the company's directors of engineering in 2012. Many transhumanists saw this as 'a symbolic merger between transhumanist philosophy and the clout of major technological enterprise, wrote author Meghan O'Gieblyn in a 2017 essay, 'Ghost in the Cloud'.[15] This has accelerated investment in transhumanist technological solutions, from AGI to nanotechnology to brain–computer interfaces.

O'Gieblyn is a former evangelical Christian and former transhumanist who investigated her own path through faith – in both the Christian God, and the digital one – in her 2021 book *God, Human, Animal, Machine: Technology, Metaphor, and the Search for Meaning.* She grew to see the transhumanist endpoint of the Singularity that both Bostrom and Kurzweil described as a replacement for what she had been taught in her childhood religion. 'In hindsight, it is strange that I did not notice the resonances between [transhumanist] ideas and the promises of Christian eschatology – at least not initially,' she remarked, referring to a field of study in theology that is concerned with death, judgement, final destiny of the soul, and the end of the world. 'Like the biblical prophets, Kurzweil believed that the dead would rise, that the earth would be transformed, that humans would become immortal.'[16] All thanks to innovations in technology.

In 2022, he explained to an audience at RAADfest: 'We will do this by replacing the cell's nucleus with a nano-engineered counterpart, which would receive an upgraded DNA code from the central server.' This would fix anything physically wrong with us at the precise point where it is going wrong.

'We'll be funnier, sexier, smarter, more creative,' Kurzweil said to the audience. 'We'll be able to produce an optimised body at will. So we'll be able to run faster, longer, whatever we'd like to do – climb Mount Everest for vacation! Sit at the bottom of a pool without drowning! We'll get to a point where we're no longer dependent on the survival of our biological bodies for ourselves to survive!'

'Kurzweil claims he is a "patternist", characterizing conscious-ness as the result of biological processes', wrote O'Gieblyn. 'These patterns, which contain what we tend to think of as our identity, are currently running on physical hardware – the body – that will one day give out. But they can, at least in theory, be trans-ferred onto nonbiological substrata: supercomputers, robotic surrogates, or human clones.'[17]

In the near future, our brains will 'eventually become more than 99.9 percent nonbiological', Kurzweil promised in *Wired* in 2024.[18] We will have synthetic neurons which will update accord-ing to the instructions of a central control system that will live in the cloud. This will communicate with the digital layers of our neocortex. The digital neocortex is what Elon Musk is building with his brain–computer interface company Neuralink.

Musk explained why we will need this to an audience in January 2017 at the Future of Life Institute's Beneficial AI con-ference, seven months after Neuralink was founded. 'We're bandwidth-constrained, particularly on output', he said. 'If you want to be generous, you could say maybe it's a few hundred bits per second . . . The way we output is like we have our little meat sticks that we move very slowly, and push buttons or tap, tap a little screen.' He's referring to our fingers. 'Compare that to a computer which can communicate at the terabit level . . . We have to solve that bandwidth constraint with a direct neural interface, I think a high-bandwidth interface to the cortex, so that we can have a digital tertiary layer.'[19]

Musk's brain–computer interface company is only one stepping stone in the transhumanist endgame, in which we 'achieve a symbiosis with artificial intelligence'.[20] Digital sensors, AGI, neural integrations – they will all be used to boost our capabilities in pursuit of transhumanism's three goals: super-longevity, superintelligence, and super-happiness.

In 2014, Bostrom published the bestseller *Superintelligence: Paths, Dangers, Strategies*, which was endorsed by Musk, Altman and Bill Gates.[21] Superintelligence, he wrote, is 'any intellect that greatly exceeds the cognitive performance of humans in virtually all domains of interest.'[22]

For starters, by seeing patterns that humans are incapable of seeing, it could help cure disease, increase food production, generate affordable energy and protect the environment, and distribute these achievements evenly, regardless of class, socioeconomic status, or geographical location. It could also solve ageing. By leveraging its own logic, it could invent new technologies and new therapies, and new strategies for extending our lives. It could. Conversely, it could dismiss humanity as insignificant, decide that this whole exercise is pointless, and instead prioritise other endeavours, perhaps even rendering us obsolete. You've seen *The Terminator*, right?

Transhumanists today wield enormous power in Silicon Valley. Their ideas are embedded into emerging technologies at Google, Tesla, Apple and SpaceX. 'All of us already are cyborgs,' Musk said at the Future of Life Institute conference in 2017, while sitting next to Bostrom, Kurzweil, Skype co-founder Jaan Tallinn, and Hassabis, who added, 'If we do this right, it's going to be the greatest thing ever to happen to humanity.' Altman and Larry Page were also attending the conference. 'If [AI] reaches a threshold where it's as smart as the smartest, most inventive human, it really could be only a matter of days before it's smarter than the sum of humanity,' Musk continued. Today, because of the development over the last decade of even more powerful machine learning, this community of AI developers, transhumanists and techno-philosophers believes the Singularity is not if, but when.

These techno-fundamentalists have faith that they have the ability to bring about the thing that will cause our evolution, or our destruction. 'Though few transhumanists would likely admit it, their theories about the future are a secular outgrowth of Christian eschatology,' explained O'Gieblyn in 'Ghost in the Cloud.' 'What makes the transhumanist movement so seductive is that it promises to restore, through science, the transcendent hopes that science itself obliterated.'

Émile P. Torres grew up in a fundamentalist evangelical Christian family in Maryland, USA. In their childhood, they believed immortality would come if they demonstrated their faith to Jesus Christ through action and thought. However, their faith preached an additional feature: they would only know if they'd done enough after the Rapture, at which point they would be judged either a believer and destined to live in a new heaven and a new earth where righteousness prevailed, or wicked and destined for eternal damnation. 'So, quite terrifying,' they say to me from their office in Hanover, Germany. We are on Zoom, and Torres, in their early forties, is wearing their trademark beanie hat and has white headphone wires dangling from their ears. They are no longer an evangelical Christian; they left their church long ago, and at the time of our call they were finishing their PhD in philosophy, studying apocalypses and fundamentalism.

'If you mention this at parties, there's really one of two responses,' Torres says about their PhD topic. 'Either people get really interested in it and have just a million questions that they want to ask you, or it sort of kills the vibe.'

I am one of the apocalypse-curious, I confess with a smile.

'The topic is dark and kind of dismal,' they demur. 'Many

people who are engaged in it tend to be, by disposition, rather cheerful.'

I take that as a compliment.

Torres tells me that the Rapture is an apocalypse scenario that eschatologists study. 'Apocalyptic tendencies go back at least to Zoroaster, two millennia before Christ,' they explained in a 2016 interview with *Indy Week*.[23] Raptures provide the faithful with guidelines for living by emphasising the importance of being prepared and vigilant, focusing on spiritual growth, and offering ultimate hope.

But Rapture stories changed with the rise of technologies in the first half of the twentieth century. The atomic bomb 'brought the possibility of secular apocalypse,' Torres explains to me, and eschatology researchers expanded its list of potential doomsday sparks. Over the last seventy years, the list of possible catalysts has grown to include biotechnology, nanotechnology and artificial intelligence. These have been folded into a category that describes any threat that could permanently and drastically limit humanity's long-term potential. They have come to be known as 'existential risks'. That is what Torres is an expert in.[24]

Bostrom has been writing about existential risks for more than two decades. He defines them as something 'that threatens the premature extinction of Earth-originating intelligent life or the permanent and drastic destruction of its potential for desirable future development'.[25] They are distinct from other kinds of risks, he explained in a 2002 article, because they are a recent phenomenon: 'We have not evolved mechanisms, either biologically or culturally, for managing such risks.'[26] So the category includes nuclear war, climate change, pandemics, a meteor strike, badly programmed or flawed superintelligence, accidental misuse of nanotechnology, and other unpredictable accidents that could threaten the future of humanity.*

Existential Risk Studies emerged as a research field after the Second World War in science-related fields, but it wasn't until the publication of Bostrom's 2002 article that Existential Risk Studies unified into its own discipline.[27] Thereafter, he founded his institute in 2005, and others followed, including the University of Cambridge's Centre for the Study of Existential Risk (CSER), funded in part by a $200,000 donation from Jaan Tallinn.

As Torres began to separate themselves from their religion in the early 2000s, they discovered Bostrom's and Kurzweil's writing, as well as an active online community. They fell easily into this group. What appealed most was that those existential risk conversations described human extinction events that were based on science and technology, rather than on religious faith. It was rational, they said. It was real. They renounced their faith, left their family and community behind, and went in search of the good place they were promised, from a god they could touch: technology.

'All of the same things are promised, except that rather than relying on supernatural agency we rely on our own ingenuity and rationality to create this world,' they explain to me. 'It provided the sort of hope that I and many other people were then missing: that the future could be better. We could live forever. There could be radical abundance. And we could all become superintelligent post-human beings living a state of surpassing bliss and delight.'

'To quote Elon Musk,' Torres continues, 'spread the light of

---

* Bostrom also includes 'something unforeseen' in his taxonomy of humanity-threatening events.

consciousness into the accessible universe.' Torres had discovered the digi-god.

Fuelled by the promises of the Singularity, they moved to Silicon Valley in the early 2010s to work as a research assistant on Kurzweil's 2024 book, *The Singularity Is Nearer: When We Merge With AI*. They began to publish widely about machine superintelligence, human augmentation and existential risks, and took on positions at existential risk research centres: the FHI and CSER, as well as Bostrom's US-based techno-progressive think tank the Institute for Ethics and Emerging Technologies (IEET), and the FLI. Torres was at the heart of the emerging international AGI community, spreading the good word in articles, at conferences and online.

More institutes continued to pop up around the world, such as Stanford University's Existential Risks Initiative and the University of Chicago's Existential Risk Laboratory (XLab). Torres's colleagues were philosophers, economists, legal scholars and technologists. Open Philanthropy had recommended an investment of about $64 million in 2021 for research to prepare for the possible risks of artificial intelligence.[28] In 2022, the US Congress passed the Global Catastrophic Risk Management Act. Like the Y2K apocalypse, the possibility of a post-AGI world was being taken very seriously indeed.

'If one believes that the future could contain astronomical numbers of super-enhanced posthumans in a galaxy-spanning techno-utopian paradise, then one should care about every possible event that could preclude humanity from achieving that goal,' Torres urged business, government and academia.[29]

Together, eschatologists, technologists, techno-optimists and transhumanists have built strategic plans for how to avert the worst outcomes brought about by a superintelligence – focusing specifically on addressing cognitive biases, coordinating decision making and conflict resolution, establishing early-response systems, and planning and maintaining large-scale evaluation and recovery infrastructure – and how to optimise the chaos for the good of humanity in the long term (which will be discussed in more detail in the next chapter). As for the short term, Torres argued, there would be some pain but, with proper preparation, everything would ultimately settle into a better, posthuman future.

And then something changed. Torres began reading more widely about apocalypses and how fundamentalists prepared for them. They started to see that the papers they were writing for the think tanks and the non-profits echoed the warnings of people in the past. They began to recognise the story beats: everything we understand will be destroyed; humanity will be transformed into something fundamentally different, and superior; we need to approach the end with unwavering faith; it's unavoidable, but after it passes and we are judged worthy, we will reach the other, better side. How was faith in Moore's Law any different from the faith they'd grown up in, they began to wonder. And as they watched their community's ideas take hold in Silicon Valley among VCs, engineers, and those thought leaders with evangelical zeal, they grew fearful. 'I'm very concerned moving forward as artificial intelligence becomes increasingly powerful that these people are going to become increasingly convinced that they're in an apocalyptic moment,' Torres tells me. They were concerned that the people building AGI and superintelligence would justify extreme actions against anyone trying to stop them 'Because what's at stake is astronomical amounts of value. It's literally galaxies,' they explain. It could be the end of humanity if the wrong people were building a new beginning.

The Singularity is what is known as a millenarian project: it promises a solution to an impending, insurmountable problem, delivered by an outside source. A deus ex machina. In other words, whatever uncertain contradiction or downright unfair situation we find ourselves in will be handled by someone or something else. In this case, an artificial intelligence built by people who see humans as data.

Historically, millenarianism has arisen out of social anxieties about major changes: new millennia, new technologies, new pandemics. But what does it mean if humanity is inevitably, unstoppably, almost out of habit, about to invent a superintelligence?

'A great future isn't complicated,' Sam Altman promised in an essay entitled 'Moore's Law for Everything,' which he published in 2021, more than a year before his company released the first ChatGPT to the public.

'We need technology to create more wealth,' Altman wrote, 'and policy to fairly distribute it.'[30]

Oh, Sam. If only it were that easy.

CHAPTER 9

The Most Good Place

And, for an instant, she stared directly into those soft blue eyes and   ⎣
knew, with an instinctive mammalian certainty, that the exceedingly
rich were no longer even remotely human.

William Gibson, *Ground Zero*

Eliezer Yudkowsky was a precocious, extremely intelligent child. At age eleven, he read *Great Mambo Chicken and the Transhuman Condition*, a light-hearted non-fiction book by journalist Ed Regis which investigates the scientists planning for a future of 'post-biological' people.[1] As Yudkowsky noted in an extensive autobiographical public document called 'The Meaning of Life', reading *Mambo Chicken* helped him learn 'that human civilization was heading towards a *much* better standard of living for everyone.'[2]

Later, Yudkowsky became a 'singularitarian' after reading Vernor Vinge's 1987 sci-fi collection *True Names . . . and Other Dangers*. The book introduced him to the concept of cyberspace, riffing on its possible political power. Not long after, he discovered the Extropians.

Yudkowsky was erudite, and organised. In 2000, when in his early twenties, he founded a private research organisation called the Singularity Institute for Artificial Intelligence (SIAI), receiving backing from internet entrepreneurs and husband and wife Brian and Sabine Atkins, who were also on the Extropian mailing list.

Yudkowsky was concerned AI wasn't coming along quickly, or carefully, enough, and this was the SIAI's raison d'être: as a dedicated investigator and developer, preparing to usher in AI with the greatest foresight possible. Initially, he was excited about superintelligence, but by 2002 he realised that he – and anyone who was building AI – wouldn't be able to control its internal morality, and the machine could, intentionally or not, end up killing all humanity. Yudkowsky decided instead to dedicate himself and SIAI to creating 'Friendly AI' – a superintelligence with human-friendly values. Doing this, he reasoned, he'd save the world from a potentially hostile artificial intelligence. The pathway to an AGI future had to be safe and beneficial for all.[3]

After the Atkinses' original investment, others followed suit: Peter Thiel ultimately became its single largest funder, beginning with an initial donation of $100,000 in 2006. SIAI built a portfolio of research and development projects, research fellowships, research grants, and a programme of science education. While the rest of us were grappling with how to filter out spam, Yudkowsky was already preparing for our technological endgame. I'm embarrassed to admit that I dismissed all of this as a passing fad.

Yudkowsky, ever prolific, wrote blog post after blog post on a site called Overcoming Bias with titles like 'Mysterious Answers to Mysterious Questions', 'How to Actually Change Your Mind' and 'Reductionism'. They covered topics like epistemology, AI and meta-ethics – a branch of ethics that explores the nature of morality, asking what is behind moral judgements – and how to make the most objective and unbiased decisions possible. In 2015, Yudkowsky published a series of blog posts (which served as his self-taught degree in moral philosophy) as a book called *Rationality: From AI to Zombies*.[4] This, along with his epic 660,000-word fan fiction adaptation of J. K. Rowling's wizarding bestseller, *Harry Potter and the Methods of Rationality* ('the most popular Harry Potter book you've never heard of'),[5] have become the urtexts of modern-day rationalism. Yudkowsky's definition has come to include 'reductionism, materialism, moral non-realism, utilitarianism, anti-deathism and transhumanism'.[6]

In 2006, Yudkowsky, Kurzweil and Thiel organised the first annual Singularity Summit: a 'Bay Area coming-out party for the tech-inspired philosophy called transhumanism', reported the *San Francisco Chronicle*.[7] 1,300 people turned up, including many scientists and tech industry leaders, to hear talks by Kurzweil, Bostrom, Thiel, Yudkowsky and Max More, the head of the Extropy Institute (which he had founded in 1991 and which had connected Bostrom, Kurzweil, Yudkowsky and others in this intellectual community).[8] Over the next few years, other luminaries in the AGI and anti-ageing field spoke, from Peter Diamandis and Sonia Arrison to Vernor Vinge, on topics as wide-ranging as personal genomics, brain simulation, the future of Moore's Law and long lives.

'If people stopped dying all the time, and actually had time to learn from their mistakes without dying of them, well actually that in itself would tend to fix a whole lot of problems over time', Yudkowsky told the crowd at the 2009 Singularity Summit in New York City. Next to him on stage, Thiel said, 'It's important to explore areas that people are not doing. It's certainly true of anti-ageing, artificial intelligence, just about all the topics that are being discussed here.' On Yudkowsky's other side, de Grey told the audience that the best way to make the world a better place would be to embarrass people with wild solutions.[9]

There were more than 800 people in attendance at the historic 92nd Street YMCA for the event that year.[10] De Grey and Yudkowsky grinned at Thiel, their most generous benefactor at that time. The work they were doing was beginning to take hold of people's imaginations.

In 2012, Thiel and Yudkowsky sold the Summit to Kurzweil and Diamandis at Singularity University, a futurism business institute in the heart of Silicon Valley. Yudkowsky renamed the SIAI to the Machine Intelligence Research Institute (MIRI). Over the ensuing years, he has pulled in more than $8 million from Open Philanthropy, $6.1 million from Thiel, $1 million from Tallinn, and $4.4 million in cryptocurrency from crypto billionaire Vitalik Buterin. And while other AGI think tanks have since cropped up around the world, MIRI is one of the best financed and longest running.

And so Yudkowsky is a celebrity in this community, particularly because after his time blogging at Overcoming Bias he went on to create the largest online resource for AGI and super-intelligence fanciers: Less Wrong. Less Wrong is where rationalists get to hang out with Yudkowsky and contemplate how to improve human decision making across all areas of their life – from deciding what job to do to how much money to give to charity, to ways to proactively act that will make the fallout from existential risks less devastating.

Over the years, Less Wrong and other rationalist blogs have attracted entrepreneurs and pop stars alike; Elon Musk and singer Grimes sparked their romance after she made a joke about a Less Wrong thought experiment on Twitter. Sam Altman described the sister rationalist blog Slate Star Codex as essential reading for 'the people inventing the future'.[11] Yudkowsky claimed to have introduced two of the co-founders of Google's DeepMind at Thiel's San Francisco townhouse.[12]

Rationalists use rationality like a mental decision-making
algorithm, translating it into the movement's moral code. And
because it's so widespread among this influential group of
future-architects who believe that life extension is on the cards,
we need to know what is driving them. Some of the things
they debate have been aligned with eugenics – there have been
exposés claiming that they connect race and IQ, or that they
believe biological differences are the reason for women's historical under-representation in STEM fields.[13] But this anything-
is-up-for-debate approach, rationalists argue, is in the pursuit of
a life free from bias. So what *do* they believe, and is their world-
view something we should integrate into decision making of
consequence?

First, we have to understand how they make decisions. All
decisions – big or small – are made using 'Bayesian reasoning',
a term also used to describe an inference technique in machine
learning. In a post on Less Wrong, Yudkowsky and others
define Bayes' Theorem as 'a law of probability that describes the
proper way to incorporate new evidence into prior probabilities to form an updated probability estimate', a way to continuously update one's thinking in light of new evidence.[14] This is
an attempt to eradicate uncertainty by planning for all possible
outcomes while attempting to remove subjectivity or emotion-
ality from the final outcome. So if it's a toss-up between dog
shelters and existential risk research centres, the latter would win
out, because, as Yudkowsky – who runs an existential research
centre – said in 2009, 'I don't want cute puppies to die, but a
little less money to puppy pounds and a little more money to
existential risk prevention would probably be a good decision
on humanity's part.'[15]

The idea here is that all probabilities – including those established under rational decision making – are subjective, and so

we must, morally, make decisions on the basis of rational proof. 'Nothing is just true,' wrote one commentator on a Less Wrong post.[16] 'No outcome, no matter how outlandish, should ever be assigned a probability of 0 because within the theory that would imply that no quantity of evidence, no matter how great, could ever persuade us to change our mind about it,' wrote Émile Torres.[17] Immortality may seem ridiculous, but it can't be dismissed, because we must always be open to new evidence that could persuade us otherwise. Rationalists have 'decided they were more rational than other people,' says Yudkowsky's collaborator Robin Hanson, an economist who still runs the Overcoming Bias blog.[18] Which is why they are best placed to advise us what to do next.

Second, we have to understand why they're thinking *those* thoughts. This is where immortality appears again. Not all rationalists believe we will literally live forever, but they all anticipate that technology is *likely* to continue to advance according to the Law of Accelerating Returns and will *likely* have transformative powers which *probably* will lead to radically extended lives.

One group believes superintelligence is literally going to turn us into live-forever immortals. These are the transhumanists and singularitarians. In a singularitarian's view, the time between AGI being created and superintelligence exploding is a matter of hours or days. They're not so concerned with what to do afterwards as much as how to get there. Kurzweil is their patron saint.

Transhumanists want to be able to augment their biology while we wait for post-human utopia. It's more libertarian, more political. Bostrom is their patron saint. To get to the Rapture, we should do everything we can to make it come about: throw that money into nanotech and biotech and AGI. When you take their rationality to that point, it is indistinguishable from the religious fundamentalism Torres grew up in.

The other group doesn't believe that superintelligence will bring 'surpassing bliss and delight', but that its existence in the world will fundamentally change things. After it comes to be, we will have to figure out how best to live alongside it (rather than be destroyed by it). In the meantime, we must prepare ourselves, morally, for that likelihood so our societies continue to exist and thrive. This is Yudkowsky's church. Either because they see that transition happening more slowly or because they don't think it will reach such godlike heights, they are more interested in what we do while we wait.

Within this church are two dioceses: the effective altruists (EAs) and the longtermists. They both look at how to best allocate their resources – whether money, time or influence – but the two differ on timescale.

EAs attempt to impartially assess what to invest in now by subjecting every possibility to a highly rational set of evaluations: puppies or superintelligent AI. Doing so provides an optimised pathway that takes the guesswork out of the plethora of possibilities one has in life. The idea of being the most effective with one's altruism is instinctively good. But where I have the most difficulty swallowing EA is with its definition of 'effective'.

Effectiveness is based on the 'hedonic calculus' of the eighteenth-century moral philosophy doctrine of utilitarianism. One of its key proponents, Jeremy Bentham, was a social and political reformer – part of the 'Philosophical Radicals' circle that challenged the legal status quo during the Industrial Revolution. Much like today, this was characterised by an influx of new technologies and new ways of thinking. Then, machines such as the steam engine and the spinning jenny were replacing jobs and impinging on everyday life. There was a population boom, particularly in the cities. There was an increase in literacy, commercialisation and industrialisation. Today, we have computing, the internet and AI. During both eras, disruptions were not and are not just interesting, but essential opportunities. Bentham believed that all political decisions should be made with the aim of producing the greatest amount of happiness for the greatest number of people: that the morally correct action is the one that produces the most good. Taken to its logical ends, this philosophy promotes charity over selfishness, and public happiness over private advantage, but it also raises a question: how do we do that arithmetic? And more importantly for the rationalist and EA communities, how does one do that arithmetic in a rational way – where even the most unlikely thing might still possibly come to pass?

Bentham believed that pleasure – happiness – is the only thing of intrinsic value, and we ought to do whatever we can to maximise it. If what you do makes you happy, it's right. If what you do causes you pain, it's wrong. And that applies to everyone, equally, because the more people who are experiencing value – happiness – the greater the good. In fact, your own pleasure and pain are subservient to what works for the commons. Divorce yourself from your ego and consider – rationally – what is good for society.

At first, the debates around the maximising of happiness – the Effectiveness of one's Altruism – were more akin to movie reviews than a deeply reasoned framework. EAs prefer to spend their resources on underserved or overlooked projects, and so lots of their money has also gone into longevity – traditionally an outsider's speculation racket. Open Philanthropy, which handed out more than $570 million in 2024, has awarded $8 million in grant money to Irina Conboy for her plasma research over the last decade.

But then came an intellectual turn. What if we widened the
definition of 'people' in 'the greatest amount of happiness for
the greatest number of people' to include people in the future?
If a decision will make one hundred people very happy today,
but doom a thousand people to misery tomorrow, it would not
be hard to agree that is a morally poor choice. And what, rationally, is the difference in that example between tomorrow and a
year from now? Ten years? One hundred years? Ten thousand
years? This is the idea of 'longtermism', and William MacAskill,
Bostrom's colleague at the FHI, is its patron saint: he argues
that what we do now should produce the most good for our
grandkids' grandkids' grandkids' grandkids. Future people
matter, morally, and so we must avert future catastrophes (such
as AI run amok), or change civilisation's trajectory so we can
maximise human potential in the long term.

Disgraced crypto king Sam Bankman-Fried explained it this
way in 2022: 'The things that matter most are the things that
have long-term impact on what the world will look like. There
are trillions of people who have not yet been born.'[19] With the
utilitarian hat on, each life is flattened into a single unit of
value – and therefore the trillions of future lives are more mathematically valuable than the billions of people who are alive
today. Musk's investments in SpaceX as a driver for a future of
space colonisation is one example of such thinking. After all, if
humanity will one day need to escape Earth for a more habit-
able planet, then it can be seen as rational to concentrate one's
resources on developments that may lead to millions of Martian
colonists in centuries to come, rather than housing the homeless
today.

But what's important is the calculus and where it takes you.
You *could* invest your altruism in a country in Africa with a high
proportion of disease, which would benefit both the people

living in that country today and their descendants. But in wider geopolitical contexts, perhaps you might calculate that there is a high probability that this country will remain poor because of internal conflicts and political corruption, and therefore the investment is not only futile, but morally wrong.

This line of thinking is why some in this community have been accused of flirting with eugenics. As Torres explained in a 2021 article in *Aeon*, the underlying reasoning is that people don't have inherent value. Applying Bayesian decision making to decisions about what to prioritise within the utilitarian framework would therefore give preference to the needs of those people with net-positive amounts of value. 'Elevating the fulfilment of humanity's supposed potential above all else could non-trivially increase the probability that actual people – those alive today and in the near future – suffer extreme harms, even death', they warned.[20] EA would not disagree with this. The point is that it would be morally wrong to be swayed by the emotive argument.

Bostrom's formative paper from 2002 on existential risk described the two world wars and the AIDS epidemic as 'tragic' events that may have been important to those who were immediately affected but, in the long term, 'even the worst of these catastrophes are mere ripples on the surface of the great sea of life'.[21] Alleviating the suffering caused by global poverty, or addressing the impending climate catastrophe and thereby alleviating the suffering of people in the Global South, are considered 'feel-good projects of suboptimal efficacy', wrote Bostrom in 2012,[22] and we instead should focus on addressing potentially misaligned AI, or colonising space.[23]

In practice – at least, according to Bankman-Fried – longtermism's moral logic justified stealing $8 billion from his customers. He believed in the EA pillar of 'earn to give', which means accumulating as much wealth as possible in order to invest it most effectively. He used his crypto exchange, FTX Future Fund – which was advised by MacAskill – to identify which projects would be best aligned with his mission; these were mostly AI advancement, 'space governance', and averting existential risk. What came out of the collapse of FTX, and the investigation into Bankman-Fried, is that the earn-to-give approach consolidates unchecked money and power. Bankman-Fried stated publicly that he had a moral obligation to make double-or-nothing bets because, although it was ultimately likely to lead to financial ruin, there was a small chance that those bets would pay off, and the winnings could be used to 'save' humanity.[24] He is currently serving twenty-five years in prison for wire fraud (among other financial crimes) for doing what he reasoned was best for the future.

This 'earn to give' loophole – where anything you do today can be morally justified if it will lead to an increase in happiness for future generations – is the perfect philosophical framework for the authoritarian. Not only would it be morally correct for a tyrant to enslave a population today, if their wider project promised greater happiness for a greater number of people in the future, but it follows that to oppose the tyrant would be morally evil, as would anything that potentially lowers that future population number: according to this line of thought, abortion is bad not just because of the loss of the possibility of one potential child, but because of the loss of possible untold millions of their descendants. Conversely, the pro-natalist views of Elon Musk, who is striving to have as many children as possible, might be considered morally superior because additional millions of descendants have now been made more probable. Even today, raising one hundred slightly unhappy children produces more happiness in total than just the one blissful child.

For the mathematically inclined, this form of rationality can be very attractive. To be able to attach a measurement to the human experience means equations, comparisons and solutions which can be – rationally – extrapolated to something far, far distant in the future, and also, with enough imagination, used to justify basically anything you want to do, with the explanation that once you have done it an infinite number of people sometime in the future will be better off. And *this* is where Bankman-Fried and others have fallen victim to the numbers.

The benefits of technology that will lead to healthier, radically extended lives – whether by superhuman AI or medical advances – are neither felt equally, nor at the same time for everyone. But this is what feeds the mindsets of our immortalists. They believe they've found a way to roll out eternity for the greatest good. It's superintelligence. It's accelerating technology. It's the digi-god. Can we fault them for dreaming big, or for trying?

Here, immortality would be twofold: literal (thanks to having accelerated the development of superintelligent AI) and figurative (their legacy in the world from having created a human-aligned, beneficial superintelligent AI). What they choose to do with their altruism will have a measurable impact on directing the future as *they* calculate it should be. And *they* are very influential.

Today, as a recovering transhumanist and longtermist, Torres isn't optimistic about the future. 'I would like to live forever,' they tell me, but they believe the rich and powerful will exploit the technologies and will ultimately cause harm to the rest of humanity. They believe the future the immortalists imagine is 'impoverished' as power is consolidated around money and perceptions of value, rather than the greater good.

'There is virtually zero reference to perspectives on what the future could be outside of a very narrow white, western view,' they tell me. 'There's just no reference to indigenous perspectives on the future. There's no discussion of what the two billion plus Muslims in the world might want,' they say. Other groups they fear are also left out of conversations about the future: people with disabilities, queer people. As Torres watched these twenty-first-century versions of rationalism and utilitarianism increase their influence in business, government and policy, as resources were funnelled towards issues that would make the most impact on hypothetical generations hence, their concerns grew.

Torres turned to the community to talk this out, believing it would be good fodder for open conversation and mutual sense-making. In 2019, they posted a provocative statement on Facebook: 'I am increasingly convinced that the EA longtermist ideology is one of the most dangerous views on the contemporary marketplace of axiological ideas. I worry that, if widely accepted, the result would be a moral catastrophe.' They received an immediate and opposite reaction to that open conversation.

'Very quickly after I went public with my criticisms, 50 per cent of the research community stopped talking to me, if not completely blocked me.' They were doxxed. They received death threats.

They continued their investigations, exposing historical debates pointing to noxious, eugenicist ideas circulated by Bostrom and others in Extropian chat rooms, and at Bostrom's FHI.[25]

Torres was not alone in experiencing problems with their community. In 2021, philosopher Carla Zoe Cremer from the FHI and her co-author Luke Kemp from the CSER wrote a paper critiquing the methodology that has been used to evaluate existential risk, which feeds into charitable outcomes in both the near and the long term. Once again, they imagined this would be helpful. But after it was published, they were contacted by senior scholars warning that 'any critique of central figures in EA would result in an inability to secure funding from EA sources, such as Open Philanthropy.' Surprised, they concluded that 'any field that operates under such a chilling effect is neither free nor fair.'[26]

'Critics like myself are not just a cause for annoyance … but profoundly immoral people,' explains Torres. 'Not just hindrances, but moral enemies. It's deeply disappointing and very surprising.' They have once again found themselves on the outside of a closed group that believes the end is nigh.

That other fractures, too, have come to the fore among the rationalists is perhaps unsurprising – after all, what was once a moral thought experiment has started to feel like a reality.

In May 2023, a who's who of tech, from Altman to Gates, Tallinn to Moskovitz, signed a joint statement about their concerns over AI that read, 'Mitigating the risk of extinction from AI should be a global priority alongside other societal-scale risks such as pandemics and nuclear war.'[27] But the question of how this should be done is the talk of the town. 'There is no reason to expect a generic AI to be motivated by love or hate or pride or other such common human sentiments,' Bostrom wrote in *Superintelligence*. 'Thinking of a superintelligent AI as smart in the sense that a scientific genius is smart compared with the average human being, it might be closer to the mark to think of such an AI as smart in the sense that an average human being is smart compared with a beetle or a worm.' How, therefore, can the people creating these machines be sure that they won't accidentally destroy us?

Yudkowsky and Bostrom believe in 'AI safety,' urging government and tech developers to slow down so we can be sure AGI is aligned with our human goals. Be aware of the false utopia, Bostrom warned in 2003: 'We need to be careful about what we wish for from a superintelligence, because we might get it.'[28] 'Progress in AI capabilities is running vastly, vastly ahead of progress in AI alignment or even progress in understanding what the hell is going on inside those systems,' cautioned Yudkowsky in *Time* in 2023.[29] And while no one wishes for the opposite, there are those who are taking a more laissez-faire, everything-will-come-out-in-the-wash approach. They lean towards accelerating development in order to have geopolitical and ideological control over it. The people on this team include Thiel, who's made a turn from his earliest days at Yudkowsky's side, and now feels the potential existential risks are overblown and economic opportunity should take centre stage;[30] Altman, who's concerned enough about alignment to sign the risk register, yet aims for OpenAI to achieve a moral and technological advantage over competitor countries and companies who he believes do not have 'beneficial' AGI as their goal; and Marc Andreessen, who wrote the 'Techno-Optimist Manifesto' in 2023, in which he made clear that his motivation is based on both market dominance and the promise of Moore's Law: 'We believe in accelerationism – the conscious and deliberate propulsion of technological development – to ensure the fulfillment of the Law of Accelerating Returns. To ensure the techno-capital upward spiral continues forever.'[31]

All sides are now raging at each other in hyperboles, becoming the fundamentalist extremists Torres predicted: these are the apocalypse-watchers. Yudkowsky has even famously said that, unless we stop development and make alignment the top priority, 'we are all going to die.'[32] Conversely, Andreessen has said, 'We believe any deceleration of AI will cost lives. Deaths that were preventable by the AI that was prevented from existing is a form of murder.'[33] This is a mildly more palatable version of a thought experiment that was published on Less Wrong in 2010 called Roko's Basilisk. Its author, Roko, argued that a sufficiently powerful AI would have incentive to punish anyone who knew of its potential existence, but did nothing to contribute to its development. Even hearing about it would justify such punishment, because you'd now have imagined it. Like the best kind of ghost story, this one stuck, and made it into Grimes's song 'Flesh Without Blood'.

Yudkowsky banned the post as part of a Less Wrong policy to stop 'information hazards'. In 2023, he proposed that AI development be stopped. 'Shut it down', he wrote. Or, 'be willing to destroy a rogue datacenter by airstrike'.[34]

Both the inevitability of the Law of Accelerating Returns, and the rationale to protect a far-flung future from human extinction at the hands of the AI apocalypse, are driving the immortalists, and neither side of their internal dispute is grounded in the reality we face today. It assumes that superintelligence is a given, that radical life extension is a must. There's no alternative. This spat diverts attention away from the other, more immediate, things that need attention: climate change, inequality, health discrepancies, and the rise of authoritarianism. It is an inference about resource allocation by people who are invested in the outcome. What about a world in which superintelligence doesn't happen? What if instead we bumble along as we have for millennia, gradually getting healthier and living for marginally longer? What happens then?

PART IV

Resource Allocation

CHAPTER 10

The Future Is Not Evenly Distributed

Of all the forms of inequality, injustice in health is the most shocking and inhuman.

Martin Luther King Jr[1]

It's Saturday morning in southern Georgia, and Molly Nadell and her husband Kris are getting ready to go into town and stand in line at the plasma clinic.

When I speak with her in early 2024 about why she donates plasma, Molly tells me that she never thought she'd spend her Saturdays and Mondays waiting two hours or more to do this to make money. Over the last few years, she and Kris, who live in their RV with their two children, and work at campgrounds, have been in need of extra income to support their family.

A fellow campground worker told her that plasma donation could help in a pinch. In Georgia, donors can make between $30 and $70 each trip depending on the centre, how often you donate, and what compensation and incentives are on offer.[2] Compared with other types of donation, this is little: sperm donors at one clinic in Georgia can make up to $175 per trip.[3]

The family receives Georgia state-funded health insurance,

a minimum coverage that allows them access to only the most basic care. They do not have access to a regular doctor, which increases their risk of premature death by 10 per cent.[4] If they had no health insurance at all, which was the case for 26 million people in the US in 2023[5] – mostly working age, low-income families, people of colour or people with chronic illnesses or disabilities – that figure would rise to 40 per cent,[6] and the number of uninsured is likely to increase under Trump's second administration.[7] Molly and Kris can't afford to pay for insurance premiums with plasma: between them, they only make enough to cover a month's rent at the RV ground if they each donate twice a week.

The US is the largest blood plasma exporter in the world, producing 70 per cent of global stock,[8] and the US plasma market has become one of the biggest money-makers for people strapped for cash. Even a decade ago, plasma was being described as 'the lifeblood of America's extremely poor.'[9]

This market is well-established: donation centres advertise at bus stops and on the sides of trash cans, and their facilities are predominantly in economically deprived neighbourhoods. Research from the Center for Health Research and Policy at Case Western Reserve University in 2018 found that the most frequent thing the participants bought using the compensation they received from the donation centres was food. Nearly half of their respondents used it to pay rent.[10]

Most centres are open to everyone, but eligibility varies. Donors should be over eighteen and meet basic health screening requirements. Some centres specialise in 'young' plasma, and send blood collection buses to colleges, hold competitions for high school students and offer incentives, such as movie tickets and amusement park passes. These donors can be as young as sixteen (with a guardian's permission).

This is not consistent income. Prices change every week and tend to go down as you donate more. If you fail a basic prick test because your blood count is low or if there are too many lipids in your blood, you won't get paid. How much you weigh, your blood type, your age, and how often you donate will affect your compensation – donation centres cap contributions to twice per week or have a top-end limit per year, but for those who depend on plasma donation to cobble together a liveable income, there are workarounds. Naturally, the internet is full of hacks to get the most from your fluids: 'Earn $1,000 a month donating plasma!' says one YouTuber, who talks about bouncing between donation centres so you don't get capped at any single facility, how to stay hydrated, and keeping away from fried foods the day before. On Reddit, there are recipes for iron-rich smoothies, debates about which brand of ground beef is the leanest, and tips for which companies compensate the best. They call it 'plassing.'

This is a young person's game: the older you get, the less demand for your plasma. There is an assumption that as you age, your body is less efficient at metabolising the fats that find their way into the bloodstream. Kris and Molly are in their mid-forties. Their pay-per-litre is already going down.

Donation centres sell their stock to institutions, hospitals and private clinics for treatments. And while most plasma does go towards saving lives around the world, when I tell Molly that some people are using plasma as a means to increase their life expectancy, she is enraged. Far from the elixir of life, she and Kris both agree that plasma looks like a bucket of toxins and fat.

'If billionaires want my plasma, they need to pay me a heck of a lot more money,' she says. She sends me photos of her badly bruised arm from her last two donations. 'The last time, I couldn't even donate and it was so upsetting,' she says. 'I felt like I did everything right by drinking a lot of water and eating healthy food. I wanted to help people too.'

Kris is also appalled. 'It's not playing by the rules,' he says. Kris believes people aren't meant to live forever, and Molly agrees. 'For somebody who doesn't want to die, it sounds like those people aren't at peace with themselves.'

∞

'For those pursuing a healthier and longer life, a customized health protocol is a must,' began a longevity newsletter that arrived in my inbox on one Friday in 2024. They're right, said a friend who works in medicine. There's a reason Queen Elizabeth II lived until she was ninety-six: she had access to precision medicine.

Precision medicine tailors treatment plans for preventing, diagnosing and treating illness based on an individual's genetic profile. The premise is that everyone's genome has a unique identity that illuminates how your body – and your body alone – will respond to a condition and its therapies. It is a window into your internal universe, a manual for your personal machine. If it sounds amazing, that's because it is. Although humans are about 99.1 per cent identical, the remaining 0.9 per cent of genetic variability makes us different, and impacts which conditions are likely to develop. With your molecular underclothes on display, you can be treated in a way that is just as peculiar as you.

All of this is made possible by 'Genome-Wide Association Studies' (GWAS). These use huge genetic datasets from public and private sources, such as over-the-counter DNA tests like 23andMe or Ancestry.com, to map genetic variants associated with conditions. If you have a genetic readout from, say,

23andMe, this can be compared with the results of these studies, and any overlap can be added to your medical record.

But to be fully mapped for full personalisation, you need to know more than your DNA; you also need readouts of your epigenome, your transcriptome, your proteome, your metabolome, and your pharmacogenome. All of those '-omes' (studied by a field known as '-omics') are entire sets of biological molecules: your metabolome is your metabolic molecular profile; your proteome is how your body uniquely produces and uses proteins; your pharmacogenome is how you process drugs; and your transcriptome captures your genetic traits, such as eye colour. Once these tests are combined with medical history, bloodwork and lifestyle data, your data-self is ready to be read.

This is an exciting and at times life-saving opportunity for people who are in the midst of fighting a serious medical situation. So far, the majority of investment in this emerging health strategy has been in researching red-letter conditions such as cancer and cystic fibrosis, but it's trickled down to such a degree that 10 per cent of drugs approved by the FDA for cardiovascular illness, cancer and other disorders carry pharmacogenomic information;[11] and in 2023, more than one-third of all drug approvals were for personalised medicines for different genomes.[12]

This application is not what I'm referring to, though. I'm talking about immortalists using personalised medicine as a tool to foresee potential problems that *could* prevent someone who doesn't want to die from living their longest, healthiest life. They see it more like a crystal ball.

The first GWAS was published in 2005, but it was the UK's Wellcome Trust Case Control Consortium (WTCCC) that set the standard with a large-scale study in 2007: 'With their pivotal paper, the WTCCC demonstrated that combining forces (large sample sizes), a rigorous study design (discovery and replication stages), and stringent criteria (multiple testing corrected significance level) were needed for reproducible discoveries,' an article reflected in 2020.[13] Ever more powerful technology is creating more connections; using AI machine learning to bring GWAS hits to the surface is like throwing a fishing line into the water and only reeling in big ones. Biotech companies are rushing to patent their next sure thing and longevity researchers like de Grey are eyeing up this fishing hole too; they're hoping that AI models will identify ageing biomarkers so they can accurately predict how a person might age.

$$\infty$$

Two billion people are expected to live healthier lives by 2030 than they did in 2018, according to the World Health Organization's projections. Over the last twenty-five years, the probability of dying of a chronic disease – rather than, for example, in a car crash or from the flu – has consistently decreased globally across both regions and socioeconomic status (SES) because of advances in public health, genetics, nutrition, antibiotics and vaccines, as well as the technology doctors can now use to peer into our bodies.[14]

Precision medicine is a new field, so it's also an expensive one, running into the hundreds of thousands of dollars.[15] Personalised healthcare for people folding it into a wellness or longevity protocol is available from private spas and clinics; the *Wall Street Journal* reports that there is no official count of longevity clinics in the US, but 'estimates range from roughly 50 to 800', and investment in this market more than doubled between 2021 and 2022, from $27 million to $57 million globally.[16] By

2025, investment in clinics outpaced investment in personalised medicine.[17] At the prestigious HOOKE clinic in London, precision medicine procedures range from £51,000 ($70,000) per year for a 'Healthspan+' membership to £6,900 ($9,400) for a half-day 'Investigation' package, which, according to their website, is 'Your personal in-depth journey; redefining ageing and optimising your health' and includes 'medical assessments, diagnostic screening, fitness, nutrition, cognitive function and stress resilience.'[18] At Clinique La Prairie in the Swiss Riviera, which has been providing longevity services to the ultrawealthy for ninety-four years, packages range from $55,000 per week for 'Revitalisation Premium' to a three-day medical check-up for $24,000.[19] The middle ground – spas with much more affordable, off-the-shelf services like IV drips or full-body scans – is the sector that is growing the fastest. Yet personalised medicine treatments for life extension are clearly investments that are unavailable to most.

Adding years to life will not be the great leveller many enthusiastic entrepreneurs sell. There is a disparity in health outcomes between the rich and the poor. Across many research studies and contexts, SES is a powerful determinant of health: the higher your SES, the longer you are projected to live compared with those in lower SES brackets. This is to do with discrepancies in social determinants of health, which the WHO describes as 'the conditions in which people are born, grow, live, work and age, and people's access to power, money and resources.'[20] It doesn't matter if the universe dropped you into a low-, middle- or high-income country, either, or if you have universal healthcare or live in a place where you must pay for your treatment; if you're poor, life expectancy is anticipated to be lower than if you're rich.[21]

One of these social determinants is where you live, physically. In 2023, the World Health Organization calculated the life expectancy of a child born in Sierra Leone to be 61; a child born in Japan can expect to live to 85.[22] The average life expectancy at birth in low-income countries is around 62, while in high-income countries it's 81.[23] But even within countries, and within cities, demographers have seen this trend. The Life Expectancy Zip Code Calculator tells me that people who live in a fancy part of New York City have a predicted life expectancy of 85 years (more than ten years longer than the average Georgian's[24]), while people who live in a significantly less fancy part of NYC are projected to live to 74.[25] In London, every two Tube stops you travel east from Westminster represents more than one year off how long you'll live.[26]

If you live in a place with air pollution and low water treatment, you are more likely to develop chronic conditions and die sooner.[27] If you're from a country with high levels of violence, you lose an average of fourteen years of life not just because of your likelihood of experiencing violence, but because of the uncertainty of living in a violent situation.[28] Other factors include poor education, minority status, a feeling of a loss of control, and having little social capital.[29] Each additional year of education reduces mortality risk by 2 per cent, though this doesn't account for race or disability and other potential marginalisations.[30]

Around the world, out-of-pocket health expenditure has increased dramatically since 2000. More than 2 billion people globally suffer financial hardship because of healthcare costs; millions of people are being forced into poverty annually because of the exorbitant treatment prices. That's spending on basic care, not on personalised anything.

Crucially when it comes to who benefits from personalised medicine, people on low incomes, or who live in rural areas, or who live in developing countries, or who have been marginalised, are not part of the GWAS datasets that are being used by AI to create diagnostic models. Of the more than 6,000 research studies from institutes around the world that have been published in the last twenty years which identified variations in genes that are associated with diseases, between 81 and 96 per cent of the people who were analysed were of European descent.[31]

Decades of study may have only uncovered truths about a slice of our species. The building blocks for personalised medicine are derived from the data of people from Western, Educated, Industrialised, Rich and Democratic (WEIRD) societies. 'Members of WEIRD societies, including young children, are among the least representative populations one could find for generalizing about humans,' wrote anthropologist Joseph Henrich and his colleagues in 2010.[32] Researchers argue that there are policies and barriers to recruitment, but there are also trust issues that come with recruiting a representative sample – with data protection, with the scientific body, and with governments themselves.[33] Yet if people not represented in the datasets have variations and interactions of genes and mutations that cause disease, we don't know about them. Personalised medicine is not only expensive and inaccessible; it might not even be relevant.

The social determinants of health won't be solved by increasing the number of years you live. Longevity will only amplify the inequalities that already exist for reasons that have nothing to do with wellbeing. And yet, it is a moral responsibility and a commercial opportunity to get these treatments to everyone. And this is the start of the political story of longevity: allocating resources for the future.

CHAPTER 11

The Longevity Alliance

Aging is not 'lost youth' but a new stage of opportunity and strength.

Betty Friedan

I first met Celine Halioua in 2022, soon after the entrepreneur had been pictured on the cover of *Wired*. She was moving offices into the Mission District of San Francisco; she'd got a lease in an airy two-storey office space that oozed hipster cool and smelled like coffee.

Her company is Loyal for Dogs, funded by Shernaz Daver at Khosla Ventures. And Halioua is on target to bring to market the very first drug that will extend lifespan, in any species.

'We have a very unique relationship with dogs,' Halioua explains to me while office pups bark in the background. 'We've co-evolved with them. We've shared an environment with them for tens of thousands of years. And importantly in every disease, but especially ageing, things like diet, things like environmental factors are very, very relevant. Dogs develop the same age-related diseases we do at approximately the same time in their lifespan,' she tells me. 'So, cancer is the biggest killer of dogs. Dogs will develop osteoarthritis. They develop dementia. It's thought they develop neurodegenerative disorders like Parkinson's and Alzheimer's.' If Halioua's dog–human thesis is correct, the biology should also be broadly applicable to wide swathes of the human population.

But dogs biologically age far more quickly than humans, so the effect of Loyal's treatment can be observed, recorded, tested and ultimately verified in a fraction of the time. 'If a big dog is getting sick and dying from early age-related disease at, you know, age seven, eight, nine, he's going grey at age four, he's getting a limp at age five. And the rate of ageing is so high, you can tell if a drug is impacting that in about six to twelve months. In six to twelve months, you're not gonna see anything in a person, Halioua explains. 'I couldn't raise the money probably to fund a twenty-year human study, but I sure as hell can raise the money to fund a however-many-year-long dog study.'

On average it costs $173 million (in 2018 dollars) to develop a drug. Once that drug has been developed, it takes an average of twelve years from FDA application to approval. The more complex or newer treatments can take up to thirty years to come to market.[1] Halioua's company is able to run its work at hyperspeed. With the $127 million they had raised by 2024, they began the trial of LOY-002, a drug for senior dogs weighing at least 14 pounds that targets age-associated metabolic dysfunction, reducing the insulin-like growth factor IGF-1 that Cynthia Kenyon first observed in tiny *C. elegans*. In February 2025, the FDA accepted their Reasonable Expectation of Effectiveness, which was a step closer to their conditional approval, determining that 'Loyal's data provides reasonable expectation of LOY-002's intended effect of extending canine lifespan, marking a significant milestone in the company's pursuit of conditional approval.'[2] The company hopes to begin producing the first anti-ageing drug, for dogs, in 2025.

Halioua's aim is to establish a pathway to help people too. If Loyal's treatment is approved by the FDA and gets to market, it would mean the regulator had rubber-stamped the first treatment for ageing in the world. And that precedent – that a drug can be used to address the underlying biological process of ageing – is set, regardless which species it was tested in.

∞

Nir Barzilai has been trying to do the same thing, but for people. Since 2015, the geneticist and gerontologist has struggled to secure the funding necessary. He's tried to reduce the financial and time burden by using a generic drug – metformin – that he believes can treat ageing.

'I just came from a really important meeting, the Aging Research and Drug Discovery meeting in Copenhagen,' says Barzilai on the phone to me in 2024. 'What really shocked me was, there were like fifty people who wanted to have a selfie with me. I don't understand it. Why?'

Among his affiliations are the American Federation for Aging Research, the Academy of Health & Lifespan Research, and the Albert Einstein College of Medicine in New York City. Barzilai is sixty-nine years old, and though he's no stranger to the media, he is bewildered by the fans. 'When I grew up, there wasn't the iPhone. Why do you need a picture of me? They're going to show their grandkids. Well, and their grandkids' grandkids.'

A cheeky little wink at an unimaginably long life: he can't help himself. Barzilai believes he's found the answer to tackling those underlying molecular causes of ageing that have been dogging geroscientists like himself and Felipe Sierra over the last fifteen years. He believes the solution to a longer, healthier life is already on the pharmacy shelf. It's just currently being prescribed for other purposes.

But this is where both Halioua and Barzilai are pushing what's been possible: regulators, supply chains and even politics aren't ready to bring long lives to the people. Halioua's approach is through the veterinary track, and Barzilai's is as the perfect outsider-insider.

On first glance, their attempts to have the FDA approve a single drug to treat 'ageing' – as in, a person's age changed because of a drug – may not seem like the profound and radical transformation that it is. But what they're suggesting is that 'age' is something that can – and should – be treated.

There's a lot of money being invested in trying to get us to believe the concept of age is something that is malleable – that the forward march of time can be slowed or even reversed. When I spoke with Sierra in 2024 about the underlying biological mechanism of ageing, he was working for the Saudi-backed Hevolution Foundation, which had between 2021 and 2024 invested $400 million into ageing research.[3] Diamandis's XPRIZE Healthspan (co-sponsored by Hevolution and others) will reward $101 million to the team who demonstrates improved healthspan,[4] and the US Advanced Research Projects Agency for Health announced a project at the end of 2024 that will develop and validate ways to measure ageing.[5]

These projects are hoping to find ways to extend, reverse and even treat the signs of ageing, as if it were a condition or a disease. But ageing is not one single thing. How does one measure it? One cannot reverse chronology, despite Bryan Johnson's $2 million-a-year experiments on himself. These projects instead aim to treat the biological mechanism of ageing, and in 2025 there is no measurement for this that is considered robust enough by the regulators anywhere in the world to be what any intervention could be tested against. The FDA has never approved a treatment for ageing because, before such a thing can exist, 'age' needs to be defined as something that can be treated. There are some tools used to provide one kind of age measure – 'biological clocks' – that can give a number based on changes in your body's DNA. They purport to tell your 'biological age'. But many epigeneticists think that these clocks lack accuracy and haven't been rigorously tested across diverse populations, age groups, or even medical conditions. As statistical correlations arising from extremely large datasets, it's not surprising that associations and relationships bubble up. Things may be associated with one another, but they may not cause one another. Biological age doesn't tell you much – not even when you're going to die.

What Barzilai is trying to do is to create a measure from a cluster of age-related diseases – such as cancer and dementia, cardiovascular disease and diabetes – to see if there might be a single drug that could reduce how often these conditions as a group, rather than individually, turn up in our bodies.

His candidate drug, metformin, or 1,1-dimethylbiguanide hydrochloride, has been used to treat diabetes since the 1950s, and is currently the most prescribed oral hypoglycaemic agent for the treatment of type 2 diabetes in western countries.[6] A 2024 study found discontinuation of metformin increased the risk of cardiovascular disease and speculated that the drug might have cardiovascular, renal and mortality benefits.[7] Researchers have also been testing it to see what happened when it was given to animals – first rodents, then larger mammals. A few others graduated their trials to humans, looking at whether people could be treated for kidney disease[8] and Alzheimer's.[9]

The geroscientist shared his screen with me to show slides of all the studies that have tested metformin's off-label, anti-ageing effects. The data he puts on the screen indicates that the drug reduces people's mortality if they have acute coronary syndrome, pre-diabetes, renal failure or frailty – which is classi-fied as 'a clinically identifiable state of diminished physiologic reserve and increased vulnerability to a broad range of adverse health outcomes' that becomes more common as we age.[10] 'Met-formin improves all the hallmarks of ageing,' Barzilai explains convincingly.

Already, most biohackers (and many geroscientists too) will tell you that it's on their list of daily supplements. But the problem is, he can't get anyone to fund his study with the amount he needs because no one – not Barzilai, nor the study's backer – will profit financially. If metformin is found to be a treatment for ageing, it won't be a money-spinner: it is a generic, widely used drug.

Around the same time I was in San Francisco with Halioua, an ambitious 26-year-old lobbyist named Dylan Livingston was working on his next white paper. Livingston, who was in tech-nology before getting his first taste of politics helping on Joe Biden's 2020 campaign, fell down the longevity rabbit hole in 2009 after listening to a talk online by Aubrey de Grey.

In it, de Grey had argued for a joined-up approach to con-quering death. First, ageing needed to be recognised as a disease, to open up funding for longevity research. Ageing also needed to be seen as 'an engineering problem'[11] to attract more funding for longevity research, so more interventions could come to market. But science on its own wouldn't be enough, de Grey said; technology on its own wouldn't either. One had to also invoke political magick to get on with living forever.

Livingston heeded this call. This was an opportunity for someone like him – 'with connections to policymakers and people in politics,' as he tells me in 2022 in Brooklyn – to create momentum in the radical life-extension revolution. Conven-iently, it was also a chance to flex his muscles in DC. He had business cards printed, made phone calls, and soon the Alliance for Longevity Initiatives was born.

A4LI launched in 2022, the year after Biden became the oldest US president (78 years, 61 days),* and the US Senate's average age (64.3 years) was the oldest in history.[12] This arguably made A4LI's political debut in Washington easier than it might have been a decade earlier, when the average age was 57.8 (then a record high) during Barack Obama's first term.[13] The aim is to pull political strings to increase lifespan in the US year on year. Livingston's message remains bullish; he has plans to be the first presidential candidate to stand on a longevity platform.

'For the first time in history, it appears aging is not an inevitable fate to which every human is consigned,' he wrote on the A4LI website. 'The ability to treat disease by intervening in the aging process is within our reach.' His optimism leans more towards de Grey than towards geroscience. Perhaps this is why his major funders come out of Silicon Valley. Relatedly, perhaps this is also because the commercial funders are looking to Washington to grease the wheels for their future profits in the longevity market.

The US is currently the world's largest biotech hub, but the Chinese government is putting its organisational effort into developing treatments at breakneck speed. According to Livingston's 2024 white paper, 'Policymakers' Guide to the Longevity Therapeutics Industry', by 2020, 'collectively, China's central, local, and provincial governments [had] invested over $100 billion in life sciences research and development' and were locking down patents at a rate that far exceeded the US.[14] 'If the United States does not take steps now to facilitate the

---

* Trump, at 78 years and 220 days, immediately broke this record when he was inaugurated for his second term in January 2025.

development of longevity therapies', Livingston wrote, 'it's likely the United States will lose ground to China.' Whoever got there first would keep the spoils; that would be a boost in the ongoing struggle to be the largest economy in the world. 'In order to continue to compete with countries like China in the global economy, it is vital that the United States remains dominant in this field.'[15] That kind of language certainly gets attention in Washington, regardless of who's in office.

In 2022, Livingston told me the first phase of his plan was to get more money requisitioned to the National Institute on Aging (NIA). He proposed a new institute at the NIH, the National Institute for Healthy Longevity and Aging Research, within which could be a Division of Biomarkers of Health, Function and Aging to '[facilitate] the approval and clinical adoption of innovative treatments targeting aging'.[16] Unfortunately, Livingston's hopes that ageing would be a government funding priority didn't come true: at the end of March 2025, under Donald Trump's second administration, the NIA lost 17 per cent of its workforce – including experts in Alzheimer's and diseases of ageing.[17] Livingston is optimistic, though, that the NIA will withstand any budget cuts, partly due to its billionaire backers.

He may have more luck with phase 2 of his plan: to change the law so drugs for ageing sail through the FDA. 'This new longevity industry is struggling because of some of the things that had been put into place in our healthcare system beforehand,' Livingston explains to me, a reference to the red tape of the regulator. 'They need advocating for because the potential of this field is so big to change the lives of so many.'

He proposes a two-pronged attack. The first is an Accelerated Approval (AA) pathway for anti-ageing treatments that target ageing as an outcome measure, like Barzilai's metformin study. The idea is to fast-track drugs that tackle ageing through human trials by testing them against what's called a surrogate endpoint. Rather than waiting to count how many study participants end up dying, this would make Barzilai's cluster of age-related diseases a proxy for age extension. It would also make the controversial biological age a proxy for healthspan. There is precedent: the AA system was created to speed up drug development largely in response to HIV/AIDS,[18] and it's been used to accelerate treatments for rare diseases, cancer, Alzheimer's and COVID-19, bringing treatments to patients much sooner on average than other drugs.[19] But is ageing an equivalent, if not greater, unmet and urgent public health issue? Does it really deserve a priority, expedited pathway to leapfrog the competition? 'Aging is the primary driver of chronic diseases that impair and kill millions of Americans each year,' Livingston claimed.*[20]

Critics of the fast-track model argue that speeding up the process raises quality concerns.[21] They point to hepatitis drugs that were approved but had to be withdrawn from the market because of unanticipated and high-risk side effects. Or cancer drugs that extend life by a few months but don't reduce suffering. Surrogate endpoints are there to generate preliminary evidence, but not to substitute for actual, clinical impact. Critics also say that fast-tracked drugs are undemocratic, as they often have high price tags, leaving the likes of Molly and Kris Nadell even further away from a longer healthspan.

Practically speaking, lobbying is never agnostic; the basis of policy is rarely rational argument, but is instead, far more often, a closed system of persuasive handshakes. A 'pernicious and systematic approach to influencing lawmakers,' according to the *Lancet*,[22]

---

* The actual primary drivers of chronic disease are smoking, poor nutrition, physical inactivity and excessive alcohol. Livingston's claim about ageing and chronic disease is unsustainable.

'a healthy part of the democratic process' – when used correctly – according to economics professor Katheryn Russ.[23] Healthcare is about 16 per cent of total lobby expenditure in the US,[24] and in the UK, between 2023 and 2024, members of the current Labour government received £500,000 from healthcare interests.[25]

It should come as no surprise that those who stand to earn the most from changes in legislation and regulation also hold the most sway in healthcare lobbying. Health product companies' contributions to lobbies eclipse physicians' and public health bodies'.[26] A4LI's sponsorship list is stacked with them.[27]

Sonia Arrison of 100 Plus Capital is a founding member of A4LI. She is the chair of the board, perfectly placed to influence and advise Livingston. She's been in this game for decades; alongside her 2011 book, *100 Plus*, she was one of the founding members of Kurzweil and Diamandis's Singularity University.

She tells me that the biggest challenge that she foresees is educating Congress on 'the root cause of all these diseases that are eventually killing everybody'.

'How do you know if a longevity therapy is actually working?' she asks. 'You can't wait until someone is 200 years old to decide if it's working or something.' And so the accelerated pathway is the answer.

Retro Biosciences is another of A4LI's biggest funders, listed as a 'Longevity Leader' on the company website, a tier worth $50,000. Joe Betts-LaCroix is its co-founder, and is also A4LI's board director. In January 2025, Betts-LaCroix told the *Financial Times* that he wants to discover and develop a drug and get it on the market 'in the 2020s', and wants to get a clinical trial started 'this year'. It is a short timeframe, so he's leveraging his business partner Sam Altman's super-powerful OpenAI machine learning model to turn cells into stem cells for therapies – a technique known as cellular reprogramming. It's called GPT-4b, and it is a bioengineering AI that was built specifically for Retro Biosciences. Betts-LaCroix and Altman intend to translate the insights into drugs to tackle Alzheimer's. In January 2025, they started a funding round for $1 billion to increase human lifespan by a decade.[28] An advanced pathway to treatment would suit them – and their investors – just fine.

In theory, if a treatment gets approval from a regulator, and if it's sold to a big pharma company to manufacture, everyone in the investment bloodline is rewarded – especially the early shareholders. 'What you want in your industry is to make the market size bigger, and not worry about it,' Daver tells me. The bigger the pie is, the more money there is, the more hype it gets, and the more money comes in.

Looking down the list of A4LI backers, it becomes clear that this is a small, interconnected group of optimists. De Grey is associated with two of the funding organisations – Methuselah and his new start-up, LEV Foundation. Khosla Ventures has a dizzying number of biotech companies in their portfolio,[29] and many of these are the supporters, strategic partners and longevity leaders behind A4LI. BioAge is another top-level sponsor. Their CEO, Kristen Fortney, is one of Retro Bioscience's advisors, and they're also part of the Khosla Ventures portfolio. The American Federation for Aging Research, financial sponsors of Nir Barzilai's metformin studies – Barzilai also happens to be A4LI's scientific director – is listed on the A4LI website as a 'Champion of the Cause', the second tier of sponsorship. In the spiderweb of longevity in Silicon Valley, everyone's in bed with everyone else.

'It might be a good thing for biotech actually, to have this founder mentality, to have somebody out there that's talking a whole lot, and putting it on this platform,' Daver says. 'Yes, I think it would be good for us to have our Elon Musk of Life Sciences.' A4LI is helping the cause for whoever that might be – de Grey, Betts-LaCroix, Barzilai, or Musk himself.

A4LI spent more than $200,000 between 2022 and 2024 lobbying, and the efforts appeared to pay off: in 2023, they announced the Longevity Science Caucus, a cross-party Congressional group in government for A4LI's cause. Before the 2024 presidential election, this was made up of Republican Gus Bilirakis of Florida, Democrat Paul Tonko of New York, Democrat Anna Eshoo of California (her district included Palo Alto, home of Silicon Valley) and Texas Republicans Dan Crenshaw of Houston and Dr Michael Burgess of Dallas. 'The longevity science industry now has a formidable group of Representatives who are advocating for policies and initiatives that have the potential to revolutionize longevity science progress,' read the press release.[30]

Among the caucus members were both allies of Donald Trump and some of Congress's more liberal House members.[31] The common ground between them was that each represented a district in which some of the largest industries were either healthcare and social assistance, or biotechnology development. They would all directly benefit from money flowing in, jobs being created, and a happier electorate. And although two of the caucus members lost their seats in the 2024 election, Livingston assured me that the initiative remains a cross-party issue. The caucus remained quiet in the early part of 2025, after the inauguration of President Trump, but was relaunched in late April 2025 in time for A4LI's US Longevity Initiatives Congressional Briefing, with a new membership roster including Tonko, Bilirakis and Crenshaw, along with three new Democrats: Ro

Khanna (Pennsylvania), Scott Peters (California), and Sam Liccardo (California), who filled Anna Eshoo's empty seat.[32]

∞

Once politics gets involved in an issue, the conversations around it tend to broaden significantly, and start to reflect how we as a society value the subject. Budgets shape opinions. Legislation shapes opinions. And this is the game Halioua, Barzilai and Livingston are playing. Now that there are big changes to the status quo being debated, the debate is more about what ageing means in society, and less about what it does to the body. How age is classified in medical terms has an impact on societal ageism.

Let me explain by way of another debate about ageing that's been happening on the global stage. The International Classification of Diseases (ICD) is a 130-year-old classification system organised by the WHO that ensures that when one doctor in Petaling Jaya, Malaysia, writes 'lupus' in their notes, it means the same thing as when another doctor in Springfield, Ohio, writes 'lupus' in theirs. It is a list of causes of disease and death matched to alpha-numeric codes. In ICD-11, published in 2022, there are over 55,000 unique codes that document patient problems from 'mixed hyperlipidaemia' (5C80.2) to 'unintentionally bitten by an animal' (PA75). The ICD is published every ten or so years, which, if you read it like a story-book, presents a picture of how we understand sickness and death, and the ebbs and flows of how we grapple with and make sense of mortality.[33]

Allow me to wax lyrical for a moment. To me, long-running classification systems are poetry. Sure, they may look like a list of random words and numbers, but if you think of them like the family tree of an idea, they can explain how we got to where we are today: you have the drama, pitfalls and revolutions of the last 130 years of history.

Every time the list is updated, everything is disrupted because everyone has to adapt their reporting. It involves a process of messing up the system and then recalibrating it in the hopes of something better. The WHO therefore leaves a lot of time for everyone to pitch in on changes they'd like to see, and changes they wouldn't.

The editing process on the latest edition began in 2011. The eleventh issue has some reclassifications that represent some big cultural shifts. For example, the description of code HA60, 'transsexualism', was replaced with 'gender incongruence of adolescence and adulthood', and shifted from the subclassification 'Mental and behavioural disorders' to 'Conditions related to sexual health'. 'This reflects current knowledge that trans-related and gender diverse identities are not conditions of mental ill-health, and that classifying them as such can cause enormous stigma', wrote the WHO.[34] This was celebrated as a major win for transgender rights.[35]

The words in the ICD (which is translated into ten languages) – and more specifically, the official descriptions – have consequences, which is why, when the WHO proposed a shift of code XT9T in the 'Casualty' section from death by 'senility' to death by 'old age', the proverbial sh*t hit the fan.

This may seem like splitting hairs but by designating 'old age' as something we can die from, the ICD-11 was implicitly suggesting that 'old age' was a disease. That a person who is advanced in years is diseased purely because they've survived all the things that might otherwise have killed them for longer than someone else. Grandma died of the disease 'Old Age'. This game of semantic Twister escalated into Congressional hearings in the US, parliamentary hearings in Europe, spiteful debates in academic journals, and total conflagration in online forums.

'Old age is not a disease, but ageism is', reads a Comment in the *Lancet*,[36] arguing against the change. In *Frontiers in Genetics*, Sven Bulterijs and his colleagues argued that although ageing has historically been viewed as a 'natural, inevitable biological process', classifying it as a disease would decouple it from what they describe as 'fatalism', and would legitimise biomedical efforts to eliminate it or conditions associated with it.[37] De Grey also argued for it. But the idea that old people are 'abnormal' or that ageing is 'pathological' fast-tracks already rampant societal ageism, Víctor Manuel Mendoza-Núñez and Ana Belén Mendoza-Soto, and Kiran Rabheru and his colleagues have said.[38] The WHO, cowed, backtracked. Once they realised what they'd waded into – thinking they were doing a good thing by removing 'senility' from their system – they changed the language. In the published version of ICD-11, XT9T was listed as 'Ageing-related'.*

The FDA has approved both Barzilai's and Halioua's trials. If either of these studies gets the results the researchers hope for, the US government's drug regulatory body will have approved a treatment for age. It is the political win the immortalists are hoping for – they believe it will benefit them, and it will.

Bryan Johnson also supported A4LI when it first started, funding the organisation at its second-lowest level ($5,000), but his vision, his Blueprint wellness enterprise, is different from that of the other investors' because he is building a protocol using various combinations of supplements, not searching for a treatment to synthesise and take to market. He would make money off the credibility the industry receives from both Livingston's push in DC, as well as the success of Barzilai's and Halioua's

---

* You can still find code XT9T if you search the database for 'old age'.

clinical trials. He could then test his proprietary lifespan supplements against 'ageing'. At the moment, all he can say is that his olive oil '[promotes] longevity, [and] it supports healthy weight management, emotional well-being, and circulation', and a 'healthier, more vibrant future'. His company's wordsmiths wouldn't need to work so hard if he could send his products for clinical testing. And before the regulator has even approved the trial, he could accurately say they're in the FDA pipeline. Same goes for all the treatments on the RAADfest wellness floor. It would flatten the market, leaving consumers with very little protection against ineffective supplements and scams.

'I think those issues are here to stay and we'll always have to manage this, you know', Barzilai says to me when I ask about how he and the A4LI scientific advisory board are trying to balance the geroscience view that ageing is an indication that can be tested against, and the commercial interests that say ageing is a disease and should be treated. 'My job now as President of the Academy of Health and Lifespan Research is on one side to say how excited I am and how much progress we're making, and, in parallel, to see how much noise we get that is unbearable . . . it's charlatans, it's hope over promise.

'My biggest fear with Bryan Johnson, who takes 150 supplements that cannot be good for you, is that he dies', Barzilai explains. 'Because if he dies, it will reflect on all of us, right? They'll say, hey, Bryan Johnson didn't make it, so why do you think that you can make it, right?'

∞

This is how we got to this point in our immortality journey:

In 1992, Cynthia Kenyon identifies the life-extending

molecule in *C. elegans*, and opens a whole new line of research. Her premise – that there is an underlying molecule that causes ageing – transforms researchers like the Conboys from people who document decline into detectives investigating the thing that might reverse ageing, searching for what it might be.

From 2008, microprocessors begin to parse ever-vaster datasets collected by the wearable microsensors we strap onto ourselves in the name of optimal wellness. This generates a bigger landscape of biological and behavioural information for scientists and technologists to dig around in, and for biohackers to optimise.

Also in 2008, Google Flu Trends is the first big data public health app, promising to unearth patterns that might lead to life-saving interventions. This opens up VC wallets. Lots of new molecules become candidates for an anti-ageing solution, and AI continues to unearth even more. Google launches Calico five years later.

In 2012, Felipe Sierra brings geroscience out of the shadows, giving ageing research a sexiness that it didn't have even a decade earlier. This creates a gold rush where scientific speculators hoping to strike it rich go deep into the molecular mines.

In 2015, Nir Barzilai's metformin trial pushes the US health regulator to consider testing a drug to treat ageing.

In 2023, enough noise is made that we are on the cusp of unprecedented longevity that it becomes an issue in the US Congress.

And in 2023, the first drug to extend lifespan (in dogs) goes into clinical trial.

In the history of immortality, all this has happened in the blink of an eye, so you can understand why there's a group of people who believe the pathway to extending our biological sell-by dates has become possible through science and tech. Most of us have no idea we are on the precipice of the next great chapter. But vested interests do.

Nir Barzilai, Celine Halioua, Dylan Livingston, geroscientists, wellness biohackers, James Strole and his apostles, and the venture capitalists are all trying to break one of our most fundamental beliefs about humanity: that ageing is inevitable. What happens when that is achieved?

We are absolutely not ready, but Silicon Valley certainly is. Hell, it's already writing the rules.

PART V

Longevity Nation

CHAPTER 12

Right to Try

Scientists discover the world that exists. Engineers create the world that ✕
never was.

Theodore von Kármán

The first thing that caught my attention when I met John Perry
Barlow in 2009 was his size. Here was this man of internet myth
and legend, lyricist for the rock band the Grateful Dead, in his
black jacket and black jeans, and he was much shorter than I
imagined. The internet has a funny way of doing that.

I was speaking with Barlow for a landmark TV series for the
BBC called *Virtual Revolution*, celebrating a quarter century of
the World Wide Web, and was on a whirlwind tour of interviews
with the people responsible for its social, economic and political impact: Tim Berners-Lee, Mark Zuckerberg, Peter Thiel, Al
Gore, Bill Gates, Jeff Bezos, and so many other heavyweights
who had each put their stamp on what, at that time, was still an
optimistic, positive technological future. Barlow was a smooth,
wry ideologue, whose rugged features suited his lifestyle: when
he wasn't doing interviews or appearing onstage at conferences,
he was a rancher in Wyoming. We were in his daughter's apart-
ment in New York City's Chinatown.

From the late 1980s, digital technology's juggernaut had

ploughed through culture and society. The internet had leaped the walls of academia and had become something anyone with a dial-up modem could access. But its reach accelerated in the early 1990s: the World Wide Web was launched to the public in 1993, Microsoft released Internet Explorer with Windows 95, and in 1996 President Bill Clinton signed a reformed Telecommunications Act – the first update since 1934 – which overhauled communications policy to cope with a future enabled by the internet. That year, Barlow published arguably the first great internet-age techno-political manifesto. He made it public at the World Economic Forum at Davos in 1996. It was called 'A Declaration of the Independence of Cyberspace'.*[1]

'Governments of the Industrial World, you weary giants of flesh and steel, I come from Cyberspace, the new home of Mind,' he wrote. 'On behalf of the future, I ask you of the past to leave us alone. You are not welcome among us. You have no sovereignty where we gather.' Barlow explained to me that he was annoyed that Clinton was trying to lay down the law in what he and the other denizens of 'on-line' saw as an evolving social experiment. So yes, he said. It was intentionally confrontational.

---

* Other essays preceded Barlow's 'Declaration' – like Donna Haraway's 'Cyborg Manifesto', published in 1985 in the *Socialist Review*, Robert Stallman's 'GNU Manifesto', published in 1985 in *Dr. Dobb's Journal*, and Unabomber Ted Kaczynski's 'Industrial Society and Its Future', published in the *Washington Post* during his bombing campaign in 1995 – but those authors were not invited to speak at Davos, with its audience of global heads of state or their economic advisors, nor were the essays explicitly 'technology' manifestos. Haraway's is considered to be influential in critical feminist theory which uses the metaphor of the cyborg – rather than comment on actual machines or technological design. Kaczynski's essay is a critique of industrialisation, arguing for a return to nature. Stallman's piece is closer to Barlow's, arguing for the freedom of software itself, and inspiring the open-source movement. However, only Barlow's has influenced real-world technology policy, explicitly describing internet governance and internet policy discourse through a lens of libertarian techno-utopianism.

The internet was a distinct realm, he argued. He envisaged it as a free, self-regulated, borderless zone: 'We are creating a world that all may enter without privilege or prejudice accorded by race, economic power, military force, or station of birth,' the 'Declaration' continued. 'Your legal concepts of property, expression, identity, movement, and context do not apply to us. They are all based on matter, and there is no matter here.' These were the principles of exceptionalism and experimentation that characterised the early internet, and which took hold in pockets of subcultures of people connected by interests rather than proximity. Today, since the rise of surveillance capitalism, platform monopolies and state censorship, this could all be considered naïve, and a romantic vision of an interconnected world; we called it a global village, and a global group hug.

When I started studying internet culture in the late 1990s, the 'Declaration' was a cultural North Star for those who wanted the internet to be a global commons of free expression. It was a techno-political manifesto about freedom, governance and the philosophy of the 'Net': libertarian, idealist and utopian, and an inspiration for early developers, hackers and digital rights advocates who built the foundations of both the web we use today, and the communities we participate in. I read references in earnest academic papers to JPB, as he was sometimes known. He was painted as an oracle, and a guide, and every time I met him I could understand why: he preached radical self-reliance, networked individualism, and liberty and justice for all. The 'Declaration' reflected this: 'We must declare our virtual selves immune to your sovereignty, even as we continue to consent to your rule over our bodies. We will spread ourselves across the Planet so that no one can arrest our thoughts.'

His intellectual stomping ground and the foundation for much of the early internet's values was the WELL, or the Whole

Earth 'Lectronic Link, which had made its debut in 1985. By 1996, it was the world's longest-running and most influential intellectual online community; by contrast, AOL's million account holders were split into many smaller groups, with none equalling the WELL's more than 7,000 members.

The WELL had been started by Stewart Brand and Larry Brilliant. Brand was an artist, poet and activist; he'd been a Merry Prankster in the early 1960s, travelling on a psychedelic road trip across the US with *One Flew Over the Cuckoo's Nest* author Ken Kesey, and he later published the award-winning, libertarian-leaning *Whole Earth Catalog*. Early editions of the *Catalog*, dating from the late sixties and early seventies, listed recommended 'tools' for life (there were 136 items in the first edition, including books by Buckminster Fuller, a printing press, and a tufting tool), seeking to serve as an ideological guide for back-to-earth living, free from oversight and centralised structure. It was highly influential in early hacker communities; Steve Jobs once described it as 'Google in paperback form, thirty-five years before Google came along.'[2]

Larry Brilliant was a hippie and an epidemiologist who'd lived in communes around the world. He spent a decade in India working with a WHO team on eradicating smallpox and returned to the US in 1978, getting his master's degree in Public Health at the University of Michigan School of Public Health, where he subsequently taught. He and his wife Girija founded the Seva Foundation, a non-profit part-funded by Steve Jobs, with the aim of curing and preventing blindness.[3] While in Michigan, he also founded a company, Network Technologies International (NETI), to develop networking communication technologies that would help public health workers in the field.[4] In 1984, he approached Brand with the idea of putting items from the *Catalog* online using NETI technology.

Brand agreed in principle, but what he wanted was more than a list of objects; he wanted a place where people connected by the sensibility of the *Catalog* could build a collaborative, non-hierarchical community. And while that sounds commonplace now, it's because the WELL did it first.

Other members started influential forums such as Craigslist, and developed influential platforms such as AOL. They wrote for *Wired*, the *New York Times* and the *Wall Street Journal*. They were self-proclaimed hackers for whom, as WELL member and researcher Fred Turner reflected in 2005, 'notions of virtuality, community, and the socially transformative possibilities of technology associated with the counterculture' became central to the 'historically specific constellation of technology, information, commerce, and community' in Silicon Valley.[5] *SFGate* described the WELL as 'an icon in the development of the internet,'[6] and many of its members were partying at the annual Burning Man festival, held over seven days in the wildly inhospitable desert of Black Rock, Nevada.

Burning Man is a Silicon Valley rite of passage, and for a long time it was as culturally influential in tech as MIT, Stanford and University of California, Berkeley, were in teaching the engineers to code. Swathes of techies decamped to it every year. Larry Page and Sergey Brin had been regulars since before their Google empire, and put on free shuttle buses for their employees to get lost in the desert. When the pair were looking for a new CEO, they hired Eric Schmidt, the only candidate who'd been to Burning Man, because he was best placed to understand the culture of the company already. Other big tech names who've spent time on the 'Playa' include Jeff Bezos, who first attended the festival in 1999; Elon Musk, who first attended in 2011, at the recommendation of his brother, a long-time Burner who began attending in 1998 and now sits on its board; and Facebook co-founder Dustin Moskovitz, who has attended annually since 2013.[7] JPB has been described as its unofficial co-founder.[8]

Just like the WELL – where Burning Man had its first website before it got its own in 1996 – the festival began as a semi-anarchic community. It is entirely created by the people who go there. They have to build the city they live in, manage their scarce resources, and open their arms and minds to anyone who wants to join in, just in case they have the skills they are looking for. Burning Man is the physical manifestation of the internet, organised around Silicon Valley values, relationships and technologies. Burning Man is 'an experimental prototype for human technological possibility in the middle of nowhere that stands on the principle of radical acceptance of everyone,' TechCrunch enthused in 2014.[9] It's the 'hacker ethic' writ large: 'hackers program because programming challenges are of intrinsic interest to them,' wrote Pekka Himanen her 2001 book *The Hacker Ethic and the Spirit of the Information Age* – she says they are 'passionate' and 'joyful'.[10] According to Steven Levy in his 1984 book *Hackers: Heroes of the Computer Revolution*, they believe that crucial lessons can be learned by 'taking things apart, seeing how they work, and using this knowledge to create new and even more interesting things'. It can also be a resistance movement, Levy explained: hackers 'resent any person, physical barrier, or law that tries to keep them from doing this,' and use digital tools to challenge authority and disrupt systems.[11]

So combine this ethos with access to technologies, access to information, and a general interest in tinkering, and of course a community of hackers decided to set their sights on reverse-engineering our biology. But this took molecular biologist and entrepreneur Ellen Jorgensen by surprise. It was 2008.

'I was basically sitting at my desk at work in a company that was doing biotech, and I came across a *News of the Weird* column in my local free newspaper where I was just checking out what bands were playing locally,' Jorgensen tells me one afternoon in Brooklyn. 'One of the things that was advertised was people who were calling themselves DIYbio and trying to put labs in their closets and make glowing green yogurt and calling themselves biopunks. And I thought, well, you know, we had just been through several years of George Bush Jr as a president who didn't believe in either evolution or global warming, and it's wonderful that here we have a bunch of people who respect and like what I'm doing for a living enough to do it as a hobby,' she tells me. 'But it sounds kind of dangerous.'

Despite her concerns, her interest was piqued. Jorgensen joined the community and found it was made up of people who belonged to 'a whole new generation that hadn't been burned by the original biotech investment cave-in' that had happened when the promises made by a previous generation of biomedical start-ups couldn't be kept. The new crop were people who 'made all their money in Silicon Valley on things like Facebook,' Jorgensen explains to me.

She had waded into a group of biohackers who combined access to information made possible by the internet with a hacker ethic political programme. The people in these groups believed that opening up biology was a necessary step towards a better society. This group was trying to activate a renaissance which they believed was 'going to take place outside of "science proper", away from the universities which dominate now, and funded out-of-pocket by enthusiasts without PhDs,' entrepreneur Jason Bobe explained in 2008 on DIYbio.org,[12] the online home of the non-profit he co-founded as a resource to the early DIY biology community.

DIYbio and SynBio (Synthetic Biology) were artifacts of a larger movement spreading out from the WELL and Burning

Man. They imagined that the model of information being kept from the public was the only thing dividing 'us' from 'them'. Kaitlin Thaney, an open science advocate, describes the movement as a throwback to 'when there weren't the same levels of competition in the same access to resources'.

When was that, I ask. My own experience in science before and after the dawn of the internet was paywalled barriers to journals and studies. There were times when it felt like my degrees should have been in 'inter-library loans' rather than psychology. Thaney explained that the ivory tower was built around in 1665, with the first publication of the Royal Society's *Philosophical Transactions*. 'Imagine you are the editor, where you are putting out something in print in 1665. There's a limited amount of papers you can carry for that', she says. 'Much of that publishing model was very much around the curation of what is most novel versus good science'. The publishers had to pay for the printing costs; the institutions had to pay for the salaries and tools. Over the centuries, she explains, it became too expensive to give an investment out for free.

By the early 1990s, though, internet technology had created the digital infrastructure for 'collaboration, research as a public good, science as a public good, based on necessity', she explains, and there was an explosion in repository infrastructure at institutions; researchers were able to share information, almost like a digital library. 'Originally, these efforts were focused on access, collaboration, and the sharing of content or data and the analysis of that so people can build on one another's discoveries', she tells me.

'The competitive system relies on proprietary intellectual property, and not giving out your secrets', Jorgensen confirms, so many of the 'biopunks' online were 'kind of bragging that they were getting around that and doing stealth stuff', and publishing pictures of themselves in basements 'with tubes of glowing green stuff.' 'All that did was create a backlash in mainstream science,' Jorgensen says.

As a person who'd set up several biotech labs in her career, safety was Jorgensen's major concern. 'We went from a bunch of people who really were inspired by Silicon Valley, who usually did not have a biology degree, but had a degree in engineering or computer science, and they wanted to recreate the story of the garage computer lab, right? And the problem is that you can't directly take something that's in silicon like that and reproduce it in a field where you actually are working with living organisms. It's very, very complicated to do open-source biology.'

It's also expensive. 'If you're trying to do biotech, you need storage at three or four different temperatures. You need, like, environmental chambers to grow whatever you're going to grow.' Each piece of equipment costs thousands of dollars. But Jorgensen was intrigued by the possibilities, and in 2009 she co-founded the world's first community biology lab, Genspace, in downtown Brooklyn.

'I thought, maybe I should get my hand in there and make sure that, you know, this doesn't all go horribly wrong. So it was very kind of paternalistic of me in a way to think that because I had the training I was going to swoop in. But we [Jorgensen and her co-founders, Daniel Grushkin, Oliver Medvedik, Nurit Bar-Shai and Russell Durrett] did actually sort of do that in a way with Genspace.'

Genspace was set up so 'people of all backgrounds can learn, create, and grow with the life sciences.'[13] They run STEAM (science, technology, engineering, the arts and mathematics) courses for kids and adults, and have a membership programme for entrepreneurs and tinkerers.

Jorgensen believes the lab has been a success because it rode the wave of the synthetic biology movement – 'that you could somehow turn biology into engineering, that you could stand-ardise it', as she says.

She could spot the engineers when they came into Genspace. 'They drew circuit diagrams and took different parts in biology, and they made little symbols for them, like they do for resistors and capacitors in electrical diagrams. That's somewhat useful, but ... there are just things that aren't really analogous.' The problem was that they couldn't get hold of the scientific papers, and that's why they couldn't figure out the human body. 'The thing is, all you need to learn how to code is a computer. And you need the skills of tapping on a keyboard. Everything else is a mental skill. Whereas in biology, there are actual physical skills.

'Walking into a biology lab and expecting to be able to do sophisticated work without doing the homework is like me saying, I want to go into a kitchen and be a five-star chef. You can watch all the cooking shows you want', she laughs, 'it's not going to make you into a five-star chef.'

Jorgensen believes the power of the DIYbio movement is in educating people in how to advocate for themselves. A 2024 Ipsos Global Trends study found that 80 per cent of people in a sample of nineteen countries want more control over their health, and 69 per cent said that they look up health informa-tion rather than relying on doctors.[14] The principles of openness advocated by Brand, Brilliant and Barlow have produced a culture in which people challenge experts, and experts must get comfortable being challenged. The hacker ethos has pushed back against corporate interests; one DIYbio group of parents hacked their kids' insulin pumps so they could monitor the data in real time. Another group formed the Open Insulin Project, and with assistance from the DIYbio community members were able to make a cheaper version of insulin than was on the market. A group called the Four Thieves Vinegar Collective circumvented the artificially inflated prices placed on EpiPens by building their own using 'off-the-shelf parts'. By breaking into the EpiPens, they discovered how to optimise the technology for their needs, and made a political statement about the controversial decision by its makers to illegally increase the price of the device.[15] During the pandemic, Jorgensen was involved with a collective that invented a COVID test alternative to PCR tests.

'I've had fifteen years in the DIYbio field watching people come in and try to do stuff, and the most fruitful things that I've seen are when somebody who isn't in that silo partners with somebody who is a scientist who helps them actualise it, but who might not have come up with the idea because it was too out-of-the-box.'

DIYbio and SynBio were born of the techno-political landscape Barlow championed: exceptionalism, libertarianism and personal sovereignty. These ideals have migrated from the internet into IRL (In Real Life), weaving early techno-utopian libertarianism into the law books of sovereign territories. In the US, in healthcare, they are enacted by techno-fundamentalists and the transhumanists who are pushing a legal structure called Right to Try.

In 2017, the US government signed in the Right to Try Act, a bipartisan piece of legislation that allowed people with life-threatening conditions who are ineligible to enter into a clinical trial the right to access treatments that have only passed the safety standards of the FDA – not the tests that confirm it's actually effective in treating a condition.[16] Research has found that many of the 90 per cent of drugs that fail the FDA tests do so because they don't work – not because they're unsafe.[17] This means that a treatment could be safe but isn't delivering the outcome the researchers are testing for.

In the context of DIYbio, where anyone might discover and design a treatment, or reverse-engineer an existing one, 'the Right to Try is interesting,' says Jorgensen. The tech policy blog Technology Liberation Front ties Right to Try's origins to frustrations parents had with the FDA, which was 'foot-dragging.' 'Thanks to new technological capabilities and networking platforms, the public may increasingly enjoy a *de facto* "right to try" for many new medical devices and treatments. Technological innovation will decentralize and democratize medical decisions even when the legal status of such actions is unclear or even flatly illegal.'[18]

But, adds Jorgensen, 'how do you maintain quality control if this is a DIY system?'

This is why the regulator was set up in the first place. At the turn of the twentieth century, there were all kinds of devices and pills that purported to be anti-ageing or rejuvenation treatments. Take for example the Overbeck Rejuvenator, an electrotherapy device developed in Britain that literally shocked people with up to 12 volts to restore lost vitality by injecting electricity;[19] or Nuxated Iron, a tablet which contained a derivative of strychnine plant and was known to be responsible for at least one fatality.[20]

The Pure Food and Drug Act of 1906 put a stop to many of these, but it wasn't until the founding of the Food Drug and Insecticide Administration in 1927 (renamed the Food and Drug Administration three years later) that the US began to require that all drugs and interventions demonstrate they were safe and effective before they could be sold. Anything intended to diagnose and treat, mitigate or cure, or even prevent a disease needed three clinical trial stages. The current steps are as follows. First, a treatment has to be proven safe. Then it has to be proven effective. Finally, its safety and effectiveness need to be compared to those of existing treatments on a large scale.

The Right to Try model has the potential to transform the FDA into solely a safety regulator. Even more than approving a drug to treat ageing, this would allow the supplements in RAADfest's marketplace to be classified as treatments for ageing – with no evidence that they work.*

The FDA currently maintains that a safe but unapproved compound shouldn't be produced or released to the public because if people take it, while they may not be harming themselves directly, they are *indirectly* harming themselves as the disease they suffer from isn't being treated. The Right to Try movement argues otherwise, and counts immortalists, transhumanists and techno-fundamentalists as allies. They want to be able to try longevity treatments that have not been tested against 'age' on themselves. 'I'm all over Right to Try,' de Grey says to me. 'It is the fig leaf that allows me to pretend that I don't think Republicans are from Mars. The fact is the Republicans are the people who are much more in favour of Right to Try.'

Immortalists, and many elected Republicans in Donald Trump's administration, don't believe the FDA and other regulatory bodies are the best judge of efficacy. They like to point to the opioid epidemic in the US as an example of how, in their view, the FDA failed to protect the public. Perhaps, they argue, each person should be able to decide what to take. 'I think that at this point, when we have much better understanding of how to use noisy data to see patterns that could teach us something, any one experiment by biohackers is actually useful,' de Grey explains.

When I get back in touch with Dylan Livingston again just days before Trump returned to the Oval Office in 2025, he was

---

* They could also be classed as treatments for anything else the developers wanted, because it would be impossible to ever know what benefits they might have.

eager to tell me about the lobbying he'd done since we first spoke. 'We helped pass a law in the state of Montana in 2023 that expands eligibility under the Right to Try Act,' he says. Montana Legislature Law SB422 'removed the restriction on patients who are eligible to receive experimental medications under the Right to Try Act.' That means any healthy person who wants to try a drug that's been passed for safety can. 'We could turn Montana into sort of a medical tourism hub,' Livingston says. Montana, Colorado and Arizona – all of which have Right to Try laws on their statute books – could become hotspots where, on their biohacking whims, immortalists could declare they are immune from the sovereignty of the rest of the nation.

I spoke to Montana State Senator Ken Bogner, who worked with Livingston, to dig a little deeper. 'It's a way for people to want to come here and get the treatments that they want.' He also expects it to be a boost to the state's economy. In 2025, he proposed a bill to allow for experimental treatment centres to open in his state.

Bogner ran for US Congress as a Republican candidate in 2024. When I bring up the fact that he was endorsed by the Montana Libertarian Party, he laughs. 'I try to focus my political career and legislation on just freedom and giving people autonomy and options and the ability to make the decision for themselves. And I feel like that's what Right to Try can do ... And that's one of the reasons why they endorsed me, is 'cause they think people should have the right to make their own decisions and not the government telling 'em what they can and can't put in their bodies.' He wants to push it out across the US. Isn't it ironic that he's proposing legislation that was inspired by leftie hackers from Silicon Valley? 'Yeah, I see a connection,' he says. 'I have to think about how I message it to my community, a very much more conservative culture in rural

Montana where, you know, your counter-culture hippies might be off-putting. But it's the same type of thinking.'

This hacker ethos has morphed into real-world activism: it has opened up science and undermined the sovereignty of the establishment, and it's ultimately becoming legislation. Platform economics and techno-feudalism are combining with the libertarianism of the early web, and the people who represent it – the techno-fundamentalists and the immortalists – have moved even closer to the seats of power. It took twenty years for Silicon Valley's ideologues to get to a place where they can meaningfully act on the ideas Barlow wrote into his 'Declaration of the Independence of Cyberspace', and these powerful billionaires who are dreaming about a better, longer tomorrow are on target to disrupt the nation state.

CHAPTER 13

The Economics of a Long-lived Life

Political science is the study of the authoritative allocation of values as it is influenced by the distribution and use of power.

David Easton, political scientist

Let's imagine a best-case scenario for our slightly extended future: scientists in the here and now identify an intervention that targets ageing – such as Nir Barzilai's metformin trial or the Conboys' plasmapheresis – and it's made available to all. Our healthspan increases, and our lifespan reaches a moderate 120 years. What does all this extra life mean for how we will live it?

To answer this question, I reach out to Andrew Scott, professor of economics at London Business School, head of research policy at the Centre for Economic Policy Research, and former director at the UK's Financial Services Authority. He and other social scientists who study our changing demography say that this evolution in our healthspan is going to require dramatic adaptation to our political and social lives. On our call, he tells me that already in 2022, for the first time in history, the young can expect to become the very old. We are more likely to live to our nineties than previous generations, he explains.

'The hundred-year life is about longer working careers and

multi-stage lives,' he says. The three-stage lives we have experienced in wealthy countries since the Victorian era – education, work, retirement – will become five-stage lives: education, work, education, work, and eventually retirement. 'We are changing how we allocate time across our life course,' he says. We are already getting married later, having kids later, and even getting divorced later. Even more life will mean that jobs, relationships, locations might last for a few decades; and then the next ones could last for a few more.

Scott was in his fifties when he realised how different his life was from either his father's, when he had been fifty and was ready for retirement, or his son's, then in his twenties and taking a gap year before starting work. 'It's not just that there's a change in the age structure. We are changing how we age,' he says.

Scott decided to turn his economic forecasting skills to the business of longevity. In 2018, he co-founded the UK's Longevity Forum, a non-profit initiative committed to 'achieving longer, healthier and more fulfilled lives for as many people as possible.'[1] He's also written several books, the latest being *The Longevity Imperative: Building a Better Society for Healthier, Longer Lives* (2024). In it, he outlines what he calls an 'evergreen' agenda, by which he means making changes at both the individual and political level that will help us make the most of our longer, healthier lives. Longevity isn't about decline, he writes. Becoming chronologically older doesn't need to be either. 'We tend to underestimate the capacity of late years,' Scott says to me. 'At an individual level, we don't invest enough in our future. At a social level, we don't really have social models for conclusion.' He calls on governments to pitch in because, for them, the benefits of having a population that lives even one additional healthy year will bring in tens of millions to the economy.

The figure he bases this on is from a study he and two colleagues published in 2021.[2] They looked at how much it's worth – economically – to live longer. An individual's willingness to buy consumer goods, work, or go on holiday are based on four factors: how much they earn, interest rates, their retirement age, and how long they expect to live. If a person is healthy, and expects to live longer, they're more likely to continue or increase their spending. If they become unwell or expect to die sooner rather than later, they'll keep their pocketbooks closed. The assumption here is that health is good for the economy.

Scott used the 'Value of Statistical Life' (VSL) predictive model to calculate the monetary value of the gains from the rate at which we age. As of 2023, the average VSL for people in the US between the ages of 25 and 65 was $11.5 million, assuming an average total life expectancy of 78.9 years, and an average healthy life expectancy, free from diseases of ageing or disability, of 68.5 years. Across the entire US population, based on 2021 data, an increase of one year of life would generate $38 trillion for the economy. Use that sum to tackle the 2024 US national debt, and there'll be a whole lot left over for Bryan Johnson's Nutty Pudding. That figure has become something of a regular statistic on pitch decks when start-ups are looking for investment.

But right now, he argues, we face the real prospect of a lack of resources to enjoy our long lives. An 'ageing society' costs, at the policy level and at the individual level, he says. A 'longevity society' will not.

Transforming into a longevity society means changing how fiscal policy anticipates life expectancy. At present, Scott explains, life expectancy is static at the moment of measurement. This means that in many developed nations an 8-year-old today is predicted to die in their eighties – the same age a 70-year-old today is also predicted to die. The problem is that this figure ignores any advances that might come along in health over the next sixty years, and clearly it doesn't take into account the reality of the future strain on individual opportunity and public finances.

Just over a decade ago, in 2012, the International Monetary Fund (IMF) published a report that predicted elderly people around the world would consume a growing share of resources, from pensions to social security, to national health services.[3] They expected this would put a strain on national balance sheets, which would not have accurately considered how many older people there would be, or taken into account the costs of this population's unique needs. We know that in 2025 we do not have cures for diseases of ageing, but far more people are living chronologically longer, and therefore there are far more people who aren't in good health.

In Japan, consistently one of the world's oldest populations, where more than one in ten people is now aged eighty or older,[4] the IMF anticipated that the rate of ageing in that country specifically would stretch public finances, and would require the government to increase age-related spending on healthcare and pensions.[5] In December 2024, when the Japanese government announced its budget, this prediction came to pass: it set a record for the largest spending on social welfare for the country's elderly in its history.[6]

Treating age-related illness costs national health services around the world billions per year. In the US alone, heart disease and stroke are projected to cost $1.8 trillion by 2050;[7] the annual price of cancer care is expected to hit $246 billion by 2030,[8] and in 2022 the cost of diagnosed diabetes was $413 billion in medical costs and lost productivity.[9] Official estimates predict that by 2030 between 20 and 22 per cent of Japanese people over sixty-five will suffer from dementia.[10] Even if a treatment becomes available in the accelerated 5–10-year timeline, societies will still need to adjust for a ballooning number of cognitive and physical impairments in the meantime.

But meeting people where they are – to alleviate the strain on healthcare systems – is also expensive.

Transportation, housing, work and education all need to come together in a complex framework of systems, says Joe Coughlin when I speak with him in 2022. He's the head of MIT's AgeLab and author of *The Longevity Economy: Unlocking the World's Fastest-Growing, Most Misunderstood Market* (2017). He shares a World Bank report published in 2022 which recommends, at minimum, improving public transport, increasing accessibility to buildings, and bringing more commercial and community businesses into residential spaces. The next stage would involve updating communications, monitoring, additional human support, and technological development. This is a lot to add to a budget;[11] the World Bank didn't put a number down for how much to put aside for this kind of project, but any urban renovations would cost governments in the billions.

One way to reduce the impact on healthcare systems is to encourage people to remodel their homes to age-in-place, with single-floor living, wide doors, accessible entrances and grab bars. These are adaptations which, alongside the broader economic benefits, can have psychological benefits for individuals, for example by helping maintain independence – but these kinds of pre-emptive modifications often rely on the means of the individual and their family, Coughlin tells me. Some nations, such as Japan, Canada, New Zealand, Sweden, and several countries in the EU, offer financial subsidies, grants or loans to help people fund a remodel, but most do not, unless the older person requires it for current medical purposes. In the US in 2025, the second Trump administration sought to make sweeping, 40 per cent cuts to the federal programmes that oversee critical care for older adults and younger people with disabilities who live at home.[12]

The position Scott's Longevity Forum is taking is that longer lives will exacerbate inequality unless the state steps in to support people with lower incomes and less disposable wealth to hand. They are calling for a partnership between the public and private sectors to 'equip us for this new reality'.[13]

Scott explains what he means: during the twentieth century, the state invested in the three-stage life: education, work, retirement. In a multi-stage life, there would be additional opportunities for it to contribute: support for retraining, for example, or social support for lower-income households during periods of financial strain, such as unemployment, having children, or being out of work due to illness.

At the moment, individuals are being offered 'longevity insurance' by private firms to help them outlive their retirement savings. But an insurance policy that pays out over the decades after retirement requires disposable income. Scott says that without the benefits of a state retirement pension for lower-income workers, they are likely to end up with a much longer working stage – without the ability to retrain – and very little retirement.

By 2050, one in six people worldwide will be sixty-five or older.[14] *New Landscapes of Population Change*, a Hoover Institution report from 2022, predicted that many countries will soon have more people leaving the workforce – and collecting a pension – than entering it. This could happen in the UK by 2029.[15] Pension funds and insurance companies are trying to hedge against longevity risk – the real potential for solvency issues that could happen with even minimal increases in life expectancy.[16] Several countries, including France, the UK and China, are responding by raising the statutory retirement age.[17]

'If you can get a therapy that's going to allow you to be really, really healthy – like feel like a 30-year-old when you're seventy – then do you really get to retire at seventy and have the government pay for you? I don't think so,' Sonia Arrison says to me in her Sand Hill Road offices. 'People don't want to hear that because they had expectations growing up that they were going to be able to retire at a certain age,' she says. 'It's hard to change people's paradigm.'

And in some places, statutory retirement age raises have not gone down well: in 2023, the French government raised the age from sixty-two to sixty-four for everyone born in 1968 or after,[18] which sparked four and a half months of nationwide protests, strikes and riots. Those demographic changes we are already seeing are affecting economic security, political stability, workforces, migration, and more.[19]

In *Golden Years: How Americans Invented and Reinvented Old Age* (2024), demographer James Chappel writes that in the US there's been little change in federal support since the 1960s for elder care, nor have there been significant shifts in Social Security or Medicare. There have been investments from federal sources that have benefited older people who are healthy and vibrant, however. The result is a successful system for healthy older people's needs, but a 'catastrophically underdeveloped' system for older people who are declining.[20] In the UK, where the government's equalities framework does focus on demographic change and ageing,[21] research from 2025 by the Institute for Employment Studies (IES) for the Commission for Healthier Working Lives found that older people with health issues are more likely to be out of work than their counterparts in other European countries such as Denmark and Luxembourg.[22]

Economists also have their eyes on what could be in store for those who might be most economically at risk in the future, when today's 8-year-olds eventually reach their 100th birthdays – job shortages, housing shortages, and non-existent welfare. When I spoke with friends in their twenties, thirties and forties about living to 100, they didn't see a bright future. They pointed fingers at people in their sixties and seventies whom they saw as delaying retirement, holding on to housing, and landing them with the future bill for a climate emergency. In the Silicon Valley longevity circles, I only hear optimism: they don't talk about the downsides; they are super excited about the longer, healthier lives they believe they are helping to create.

Arrison is excited about the extended period of 'adulescence' that will give people a chance to get to know themselves better, because they won't need to be committed to one person for life, or to one job: 'With longer life expectancies, people would have the opportunity to retrain to do something more appealing to them,' she wrote in *100 Plus*. 'A person could spend thirty years in the business world, thirty years as a doctor, and an additional twenty-five years accumulating human capital through education.' Sure, if they have the income to do so. Countries that subsidise education, such as Germany or Norway, wouldn't be able to fund generations of serial retrainers. And those countries that don't prioritise re-education through supporting loans could see an even bigger inequality gap.

She's also excited about pushing reproductive health beyond the current limits of our biology. One of the companies she invests in, Gameto, is working to develop treatments to extend reproductive age. Arrison imagines 'clusters' of full siblings separated by multiple decades: multi-generational living situations that will solve the loneliness epidemic for older generations; younger-generation siblings benefiting from within-family knowledge; social capital being ever stronger. But what impact would all this have on women's bodies, and their careers?

Aubrey de Grey is more excited that longevity escape velocity will stop 110,000 people per day from dying.

'Ageing is so expensive,' he says. 'It's not just that it kills people; it's not just that it causes a huge amount of suffering and therefore there's a humanitarian imperative to make sure that this is available to everybody who's old enough to need it. But also,' he pauses, sighing wearily, 'there's a sheer mercenary economic imperative. The fact that the vast majority of the medical budget of the western world goes on the health problems of late life.

'If you've got therapies, even if those therapies are fairly expensive to deliver, that stop people from getting that way in the first place, they will pay for themselves many, many times over, really quickly. So, it will simply be economically suicidal for any country not to do the front-loading of investment that ensures that this is available to everybody.'

And what of the social and economic impact of longer lives?

'Everyone knows perfectly well that even really conservative organisations like the United Nations say that, by 2050, most of the jobs that exist today won't exist. We can totally forget about the idea of full employment or of dependency ratios or any other thing that we think we know today about how to distribute wealth equitably.'

Getting a solution over the line: that has consistently been the message I've heard from most entrepreneurs. Dealing with the rest, well, they'll leave that to someone else. Perhaps someone like Dylan Livingston?

Like most of the longevity enthusiasts I've spoken with, Livingston has no clear plans in mind for a post-age future despite it being a policy initiative of the institute where he works; A4LI is singularly focused on accelerating longevity treatments. His objective is to make people healthier, for longer.

'So, how much *do* you want to increase lifespan?' I ask him in 2023.

'I remember reading a study once that said, even if we cured all diseases, we're not going to live forever. So let's take that off the table,' he demurs. 'But radically longer is definitely on the table.'

'What's radical?' I ask.

'I would consider radical anywhere from ten to a thousand years.'

I push back: a boom in long-lived people could collapse society. Does he have any suggestions for how to tackle that? 'I'm so wrapped up in the now,' he admits. 'I'm not even thinking about those issues right now because they don't exist yet.'

Livingston is unexpectedly dismissive of longevity policy initiatives that are trying to deal with the very real, right-now issues of an older population that is currently suffering from diseases of ageing: the UK's All-Party Parliamentary Committee (APPG) for Longevity, for example. I asked him specifically what he thought of the APPG for Longevity, which was launched in 2019 by then Secretary of State for Health and Social Care Matt Hancock under Boris Johnson's government to rebrand 'the problem of ageing' as 'the opportunity of longevity'. It was tasked with setting up the roadmap for the UK government's plan for everyone to have an extra five years of healthy life by 2035. After convening several meetings and recommending more investment in the infrastructure changes that were also recommended by the above-mentioned World Bank report, in 2023 it – like many other APPGs – quietly closed its doors. Livingston said it didn't think big enough.

'We in the US are going to focus on the futuristic, cutting-edge biomedical research end of this rather than, you know, Meals on Wheels,' he says. My eyes widen. Was he suggesting diverting money away from this programme? 'Meals on Wheels and increasing the amount of ramps in public spaces are meant for a population that is not healthy, right? The whole purpose of the longevity ageing industry is to make that so that's not the case.'

And yet these services are essential for people across the lifespan, regardless of age, I suggest.

'Obviously, ramps will be needed, I'm not saying that,' he responds. 'We in the US and with this biotech caucus that we started are going to focus on the futuristic, cutting-edge biomedical research end, rather than what we can do today,' he repeats. He points to the US Senate Aging Committee. 'That's basically what they focus on,' he explains. A4LI will 'contrast' with what the UK is doing and what already exists in the Senate.

What I was hearing was ableism – discrimination in favour of able-bodied people. But Livingston's comments – though callous – were not unusual in this community of longevity entrepreneurs. Again and again I would hear about how they were 'solving' the ageing 'problem', and how that would 'fix' our 'broken' bodies.

During my master's degree, I used the social model of disability (SMD) to try to understand how the design of digital technologies could impact the lives of people with physical and mental impairments. The SMD explains how disability is constructed by the social and physical architectures of the world, which act as barriers to entry. Ramps and other aids give people across the lifespan access to independence; if these things were universal, proponents suggest, people with different abilities wouldn't be considered dis-abled.

But when the biohackers and wellness influencers describe age as a failing, it isn't ableism – it's ageism.

The WHO defines ageism as claiming someone is too old (or too young) to do something on the basis of their chronological age, rather than any measure of ability. In 2021, they reported that it 'constitutes an important, and hitherto neglected, social determinant of health. Its impact on health is on par with, if not greater than, that of racism, a form of prejudice and discrimination whose health consequences have been widely studied.'[23] Research has found that ageism directed at older people can affect their cognitive and physical capabilities,[24] as well as their mental health,[25] and it results in worse health outcomes.[26]

As with the social model of disability, we may not realise how ageist the social environment we live in is until we find ourselves personally affected by it. It is pervasive, globally, wrote Alana Officer and her colleagues in a 2020 paper. Thirty-four out of the 57 countries that they observed as part of the World Values Survey were classified as moderately or highly ageist: 'at least one in every two people included in this study had moderate or high ageist attitudes', they wrote.

So what do they (we) see? 'Older age is generally typecast as a period of frailty and inevitable decline in capacity, with the depiction of older people as a homogeneous group that is care dependent, burdensome on health and social care spending, and a hindrance to economic growth.' This stereotype, Officer and her colleagues say, is inconsistent with reality.[27]

Yet we see ageism in the media, and hear and read it in everyday language. We live by it in institutional policies, like mandatory retirement, and practices, like shortcomings in training programmes dealing with issues specific to the older population. These stereotypes are damaging. They affect how we think, feel and act.

It's 'age denialists' who are pushing a longevity agenda – they accept age, but only if it's healthy and able. Will the people who are responsible for allocating the resources in society repudiate age, as some of us live longer and healthier lives, while others don't?

CHAPTER 14

The Arms Race

You know, there are those people that say it takes over the human race.
It's really powerful stuff, AI. So let's see how it all works out.

Donald Trump, June 2024[1]

In 2025, Donald Trump returned to the Oval Office with bold ambitions to rework the US government. Behind him at his inauguration were Mark Zuckerberg of Meta, Jeff Bezos of Amazon, Sundar Pichai of Google, and Elon Musk. Also in attendance were Sam Altman and Tim Cook, CEO of Apple. These representatives of the biggest software companies in the world had thrown their financial and reputational weight behind the new administration and their presence signalled a new era in the relationship between Washington and Silicon Valley. The country was to embrace techno-optimism, and AI would be the present – not the future. But the cost has been to gut healthy-ageing initiatives, research and policy for today's older people in favour of restructuring and disrupting government so it can adopt a techno-fundamentalist agenda.

One of the people who was not standing behind President Trump on 20 January 2025 was Peter Thiel. Thiel had been a big backer of Donald Trump in 2016, sending a cheque for

$1.25 million to his campaign. In a speech at that year's Republican National Convention, Thiel admitted he wasn't a politician. 'But neither is Donald Trump,' he said. Thiel was looking for someone who would slash regulations and dismantle the state, and had been invited to speak by Trump's son, Donald Trump Jr. Trump senior, Thiel believed, would be like one of those outside-the-box thinkers he'd already invested in, such as Musk, Altman, de Grey and Yudkowsky. 'He is a builder, and it's time to rebuild America,' he said.[2] For his service, he was appointed as a key advisor on the president's transition team.[3] Several of his closest colleagues and advisors became the White House's closest advisors, with individuals considered for the top role of the FDA, positions in Strategic Communications, and positions on the National Security Council.[4] In March 2019, Thiel's former chief of staff, Michael Kratsios, was announced as the US's chief technology officer. A few months later, Thiel claimed Google was engaging in 'peculiar' behaviour with regard to its commercial relationship with the Chinese military.[5] Trump tweeted, 'Billionaire Tech Investor Peter Thiel believes Google should be investigated for treason,' promising 'The Trump Administration will take a look!'[6] This was probably the closest influence anyone from the tech community had ever enjoyed in the Oval Office, until now.

After Trump's first period in office, Thiel told *The Atlantic* that he was discouraged by the president's performance.[7] He didn't finance Trump or endorse him in 2024; he was present though, behind the scenes, and this is reportedly where the VC generally prefers to stay.

The inauguration party Thiel threw a few days before Trump was sworn in on 20 January 2025 was well attended by Silicon Valley billionaires and political heavyweights too; of course J. D. Vance, the vice president-elect, made an appearance.[8] Thiel had hired Vance to work for his Silicon Valley VC firm, Mithril Capital, had funded his Senate run in 2021, and had introduced him to Donald Trump as a potential running mate. Thiel may not have been in attendance when Trump was taking his oath of office, but his spectre was everywhere on Inauguration Day, and in the weeks after as he helped the transition team make its selections. His influence continues, with allies once again in key positions around the president, and ideas that are made easier to implement thanks to the new administration's policies.

Peter Thiel was born in West Germany in 1967, to German parents. His family emigrated to the US when he was one year old, firstly living in Cleveland, Ohio. They spent some time in South Africa before settling in the Bay Area, when Thiel was nine. He excelled in maths, was a chess champion, and went on to be the valedictorian of his high school. While at Stanford University, where he studied philosophy, he and a few other students started the conservative student newspaper *The Stanford Review* in protest of what they felt were university policies favouring diversity and multiculturalism. He followed his undergrad with a law degree, clerked for Judge James L. Edmondson of the US 11th Circuit Court of Appeals and spent some time working at New York law firm Sullivan & Cromwell, before returning to California to co-found a digital payments company he hoped would 'create the new world currency' and 'give citizens worldwide more direct control over their currencies than they've ever had before.'[9]

Thiel's entrepreneurship and investments have faithfully followed a libertarian agenda.

PayPal was created in 1998 to be 'free from all government control and dilution', Thiel wrote in his 2009 manifesto 'The Education of a Libertarian'. 'By starting a new internet business,' he continued, 'an entrepreneur may create a new world' that will 'impact and force change on the existing social and political order'[10] – which in 2023 he described as a 'senile, central-left regime'.[11]

But rather than act as a separate financial system as he had hoped, PayPal smoothed the transition to e-commerce and helped make the web a commercial, capitalist place by bringing the virtual and the physical worlds closer together.

This was not the end of his ideological adventures through business, though; he was worth $28.5 million after selling PayPal to eBay in 2002,[12] and has continued to invest in companies and people that would advance his ongoing plan to 'escape from politics in all its forms' – to find a place that governments couldn't reach, and where personal freedom was paramount.[13]

'I no longer believe that freedom and democracy are compatible,' Thiel's manifesto explains. Between 2009 and 2011, he gave $1.25 million to a project called the Seasteading Institute,[14] a nation-state-building experiment, 'to establish permanent, autonomous ocean communities to enable experimentation and innovation with diverse social, political, and legal systems'.[15] It was the vision of former Google employee Patri Friedman, who billed it as 'Burning Man meets Silicon Valley meets the water'.[16] Along with other activists, Friedman, the self-described anarcho-capitalist grandson of free-market economist Milton Friedman, imagined that seasteaders would live in a floating utopia that would 'restore the environment, enrich the poor, cure the sick, and liberate humanity from politicians'.[17] The *Guardian* described the movement as 'the ultimate Silicon Valley approach to governance'.[18] Thiel explained that 'settling the oceans' is one idea he had to 'propagate the machinery of freedom'.

Even before he began supporting the Seasteading Institute in 2009, Thiel had already set up several investment companies, started data analytics firm Palantir Technologies – its clients include the US Department of Defense, the US Intelligence Community, the CIA and, in the UK, the NHS – and created the Founders Fund, through which he has invested in companies building revolutionary technologies, favouring those – like SpaceX ($20 million in 2008) and Bitcoin ($15–$20 million in 2017) – that challenge existing institutions and promote innovation. He funded the Club for Growth, one of the largest anti-tax pressure groups in the US, with a $1 million donation in 2018. Thiel was an outlier, politically and institutionally, in the liberal Valley: 'He's a contrarian from an investing standpoint and thinks a lot about the Singularity,'[19] said Musk in 2007.

'If we want to have anything like, you know, a smooth transition as a society over the next twenty years, I think we have to actually see the question of a good Singularity as the single most important political, cultural, economic, technological question that we have,' Thiel said at the 2009 Singularity Summit. 'The faster technological progress happens on the whole, the better.'[20]

Most of his investments in the Singularity and radical life extension have been through his Thiel Foundation, a vehicle for more future-focused R&D projects. The Thiel Fellowship, under the Foundation's remit, launched in 2010, and rewards young entrepreneurs with $200,000 over two years to 'stop out of school' – as in, leave college – to be mentored by Thiel and his associates.[21] One of the first recipients was Laura Deming, a biotech wunderkind who started her career studying the biology of ageing at the age of twelve, upon joining Cynthia Kenyon's lab at the University of California as a volunteer, and who now runs one of the largest anti-ageing VC firms in the Valley, The Longevity Fund.* Other recipients have included blockchain programmers, AI developers, payments companies and health app designers. Software engineer Luke Farritor, another Fellowship beneficiary, was hired in 2025 to be part of Musk's Department of Government Efficiency (DOGE), tasked with streamlining the inner workings of the US government, and was given access to the Department of Energy's IT system in February 2025.[22]

Thiel has a vision of the future that involves personal liberty, technological acceleration, and life extension. He may have once been an outlier in the Valley, but today he's a central node in a network of powerful tech billionaires who, against all expectations, helped bring the Republican Trump back into political office. Thiel connections are once again everywhere in the 2025 Trump administration org chart: in addition to J. D. Vance (Vice President), a non-exhaustive list includes his former chief of staff, Michael Kratsios (Director, Office of Science and Technology Policy), his former colleagues at PayPal David Sacks (AI and crypto czar) and until recently Musk (DOGE), his former employee at Palantir Gregory Barbaccia (Chief Information Officer, Office of Management and Budget), and Jim O'Neill, the former head of the Thiel Foundation, board member of the Seasteading Institute, and former CEO of the SENS Research Foundation, who was nominated as the Deputy of the Department of Health and Human Services (HHS) under Robert F. Kennedy Jr – the department that oversees the NIH, CDC and FDA.[23]

* The Longevity Fund's co-founder is Loyal for Dogs' Celine Halioua.

There has for a long time been a push–pull of influence between internet entrepreneurs and the US government. The internet project was originally funded by the US Department of Defense; in her 2019 book *The Code: Silicon Valley and the Remaking of America*, historian Margaret O'Mara explains that the federal government's policies on trade, immigration and innovation helped foster growth and entrepreneurship. But once the web launched in the mid-1990s, and the bank balances of Northern California entrepreneurs began to inflate, the government was no longer the main funder, allowing ideologues like John Perry Barlow to declare their independence.

In the mid-2000s, as social network platforms grew in importance, politicians gravitated towards them; Obama's 2008 campaign made use of digital platforms, like YouTube and Twitter, which observers and analysts say was a factor that contributed to his election success.[24] When he was elected, several Valley insiders were given jobs in his administration – execs from Google, Twitter and Uber, along with biotech CEOs, took offices in DC to support the tech agenda.

The influence went in reverse too: after Obama left office in 2017, several of his staffers went to California and got policy jobs at some of the biggest names in tech. There, they joined the former leader of the UK Liberal Democrats and Deputy Prime Minister Nick Clegg, who was President of Global Affairs at Facebook (later Meta) until 2025. Valley companies hired people to liaise with Washington, and Washington hired people to liaise with the Valley. There was generally a friendly relationship, as long as Silicon Valley felt it was being kept on a very long leash. In the UK, Prime Minister David Cameron's government even briefly had an ambassador to the tech start-up incubator scene.

Throughout the 2020s, though, that leash got a little shorter.

Tech evolved away from social media into things that require ever greater amounts of computational power and energy, like AI and nation-sized bulk data. Legislation globally started to impinge on the goodwill of the executives. Under President Joe Biden, the US Federal Trade Commission came down on mergers and acquisitions – notably for technology, attempting to block Microsoft's acquisition of Activision Blizzard in 2022 – which means VCs, who are paid when a company sells or goes public, never reach pay day (this has been maintained by President Trump into 2025). Cryptocurrencies or pay-later processing firms were also designated as regulatable bodies, with responsibilities that mirrored existing offline services.

The handrails imposed on AI development also started to feel restrictive to those who wanted to dominate. They claimed that the hoops they needed to jump through to make sure that generative AI would act in our best interests – pre-deployment tests, evaluation guidance, software speed bumps – were stifling innovation, not inspiring it. DC was imposing too much oversight on the freewheeling tech industry. Many VCs in the San Francisco Bay Area, one of the most left-leaning parts of the US, quietly began to move politically to the conservative, free-market right. Former political allies in the Valley began to speak out against what they saw as a land grab that would harm their business models. Two of the biggest names in tech – Mark Zuckerberg and Marc Andreessen – jumped sides, pledging their support for Thiel's man in Washington. 'Regulatory capture' was the enemy, Andreessen wrote in his wildly accelerationist 'Techno-Optimist Manifesto'. Sam Altman, famously anti-Trump in the past, announced in early January 2025 that he had changed his mind about him; the AI mogul joined the president-elect as he announced a $500 billion AI infrastructure project called

Stargate that the administration hopes will propel the country to global AI dominance.

'The people who don't want regulation, and who are whispering in President Trump's ear, are saying, well, let's get rid of the regulations so that the existing big players and the existing business models can thrive,' Nobel Prize-winning economist Daron Acemoğlu tells me when I ask what the America First policies regarding AI mean for the rest of the world. I was interviewing him for an episode of BBC Radio 4's *The Artificial Human*.[25]

Driving the political narrative around AI is superintelligence, Acemoğlu says. 'AGI is both an assumption and it's an objective . . . a guiding light for the desire to remove regulations.'

Like conversations about 'solving' ageing, the tech accelerationist mindset is focusing too much on a fixed future rather than on a messy present. Acemoğlu believes the most dominant voices are telling stories about existential risk, which out-shouts debates about policies that will create safeguards for short-term societal impacts, or the populations who won't benefit from rapid development.

'Automation is an engine for inequality,' Acemoğlu continues. 'We're not going to get productivity benefits. In fact, we may actually do very badly on productivity, because under the false pretence that these machines are so good, we're going to automate a lot of things, and then we're going to get what? So-so technologies. All right, so, fine? But it's not revolutionary. It's like: automated customer service, or self-checkout lines. No productivity revolution. The fifth industrial revolution is not going to come from automated checkout lines, I can assure you.

'Technology benefits those people who control it – the tech companies who are the most powerful companies in the history of humanity, writing the rules of the game. We need counterweights. The extreme concentration of power is inconsistent with democracy.'

The assault on the infrastructure of the US government by DOGE is arguably a more immediate issue than AGI. The approach to solving the 'problem' of government reminds me of how technological thinking is being applied to 'solving' the human body. Both overlook the interdependencies of running a complex system, and could cause undue and unnecessary harm.

∞

2025 was supposed to be a big year for ageing, regardless of who was in the West Wing. It was going to be the year of the White House Conference on Aging, held every ten or so years to determine the policies and strategic direction for the next decade.

Previous conferences had ushered in Social Security amendments, as well as the Older Americans Act (in 1965); but it wasn't until 1995 that delegates urged thinking about 'ageing' rather than 'the aged.'

Geroscientist Felipe Sierra would normally get an invitation to the event. So would members of his former NIH institute, the NIA, and long-time ally Anthony Fauci, who is now retired from government work. When I spoke with Sierra in January 2025, he wasn't so sure that would happen. I asked him who he thought would attend.

'Well, the ones that are going to be there are the likes of Peter Thiel and all of these advocates who, really, they just have money. But they don't know anything.'

Thiel has been investing in longevity research for more than a decade now, I retort.

'I mean, I've talked with Peter Thiel,' he responds. 'I've talked with many of these people, and to an extent, maybe I am responsible for their existence. Because when geroscience came, that's when these people appeared. They thought, hey, this is true, let's focus on ageing and cell diseases. That was the whole premise of geroscience at the beginning. So, the whole industry exists because of geroscience. So they're not going to turn against it. It's just they wanted to subvert it to be faster,' he says. 'But you have to have some grown-ups as well.'

The White House Conference on Aging was cancelled in early 2025, when the Congressional budget deprioritised the event. The issue doesn't appear to be a priority for Donald Trump, who turned seventy-nine in 2025. Instead, his executive orders and administrative actions are dismantling the system that keeps older Americans healthy for longer.

In the early part of 2025, Elon Musk's DOGE implemented actions that significantly impacted research at the NIH, and regulatory oversight of the FDA. The Department cancelled leases, reduced the workforce in the FDA's Office of Digital Transformation, and terminated hundreds of research projects valued at billions of dollars. 'What they want is to destroy the system so that they can build it from scratch,' Sierra tells me after Trump's inauguration. 'There will be a shift from the more solid scientists working on the standard methods of science to those that just want to disrupt.'

And that's what is happening. 'We should reform the FDA so that it's approving drugs after their sponsors have demonstrated safety – and let people start using them, at their own risk, but not much risk of safety,' Jim O'Neill said in 2014 at a SENS conference.[26] Like Livingston and Montana State Senator Bogner, O'Neill believes efficacy should be proven *after* a drug has been rubber-stamped.

Elon Musk has also criticised the drug approval process

as being too slow. He gleefully slashed the number of federal workers responsible for developing regulations – many of whom were then asked to return.[27] While RFK Jr has begun the process of laying off 10,000 employees at the HHS, President Trump promised that the head of the agency, Dr Marty Makary, would 'cut the bureaucratic red tape' and 'make sure Americans get the Medical Cures and Treatments they deserve.'[28]

'I think this is the ripest time that we've had,' Livingston tells me. He echoes O'Neill: 'I think efficacy should also be used more as a marketing tool, maybe more than a requirement for getting the drug out there, right?'

Researchers at the National Institutes of Health aren't thrilled about this libertarian fantasy. The NIH's basic science portfolio has been slashed by DOGE; RFK Jr's priority is infectious disease. Sierra says that when the dominant source of funding is public rather than private, it changes the character of the research. It's slower, more considered, and generally obliged to pay more attention to ethical and social issues. 'Peter Thiel's allies will try to implement policies that accelerate the use of AI and things like that with unproven methodologies,' warns Sierra. He doesn't have high hopes for what the Silicon Valley appointees will offer his research. 'The problem that I see is that when those things don't yield the results that they expect, then the world will turn against it. That's the big danger to us.'

Thiel has said he wishes he'd spent more money on longevity, so perhaps this will be a rich moment for investment, but the geroscientists are wary. 'What the concern has been for a long time actually is people who know computing but don't know biology,' Sierra explains. 'They come up with ideas that are computer-derived and not biologically tested or biologically relevant. And there's too many of those proposals for anybody to take the time to check whether they're true or not. So we are inundated with things coming up from people who really don't know what they're talking about.'

De Grey shares this point of view.

'What I want governments to do is have a war on ageing,' he tells me in March 2025. 'The only way the governments of democracies are going to go anywhere near that idea is if they perceive that there is very strong public demand for it. This is why I haven't spent much of my time trying to talk directly to governments, because they might be courteous, but they won't do fuck all unless they feel there are votes in it, because they only have one goal in life, which is to get re-elected.'

His idea is to get enough people talking about a truly astonishing finding, which he hopes will come out of his lab: doubling the lifespan of a mouse. When that happens, he wants his colleagues – the small cohort of subject-matter experts who have the ears of influential people – to back him up and say, '"Fuck yeah, we are really in a completely different world now. We have broken through a century-long glass ceiling, and we suddenly have an idea of how to actually control this ageing thing."'

'Now, who's going to get funded?' asks Sierra. 'The outliers. And that is going to be the thing that I'm most concerned about: nitty-gritty, basic research – going from what you see in a worm to mammals – that's not going to get covered because it's not fast enough.'

∞

Recall Stephen Cave, the philosopher whose book explained how the human pursuit of immortality drove the development of civilisation. Today's powerful immortality seekers are doing just that: using their technology, money and influence, they are modelling political systems, regulatory regimes and legal prece-dent, fuelled by a techno-libertarian philosophy to create a new kind of autonomous political entity. All this policy change and the disruption it's causing is practice for the new civilisa-tion they want to build. But rather than settle for the borders that were drawn in the past, many immortalists imagine that the civilisation they will build will reconfigure the idea of the nation state. Today, they are pouring money and resources into a new kind of technologically enabled territory: a longev-ity Network State.

CHAPTER 15

Live Forever

I intend to live forever. So far, so good.

Steven Wright, comedian

In 2024, Bryan Johnson flew to the island of Roatán in Honduras to visit a city called Próspera for a longevity procedure. He went into the offices of the Minicircle longevity clinic and paid $25,000 for an injection of a hormone that the company claims 'may be able to treat or delay the onset of aging.'[1] It's called follistatin.

'Follistatin is a morphogenetic hormone which improves tissue composition and extends lifespan in healthy mice by 32.5%,' reads the description for the treatment on Minicircle's website. In early human trials – conducted by Minicircle – it appears to increase bone density and muscle mass, improve the sense of wellbeing, and reduce biological age. The seed funding for the company came from Peter Thiel and Sam Altman.

'I am now officially a genetically enhanced human,' Johnson announced to his 1.7 million YouTube subscribers after his treatment. One step closer, he said, 'to humanity's only objective.'[2]

This treatment is not approved by the FDA, nor by Honduras's health regulators, but 'approval' is not a prerequisite in Próspera. Próspera is a Zone for Employment and Economic

Development, or ZEDE,[3] a private city which hosts events with the tag line, 'Come Build a New City – To Make Death Optional.'[4]

A ZEDE is a form of charter city, a concept first proposed by Nobel laureate Paul Romer in a 2009 TED talk as a way for a developed country to act as a guarantor for part of a developing country. The guarantor would administer the laws and the policies within the territory, effectively taking over the day-to-day governance, and in return would have autonomy over its civil code. Romer imagined they would be a way for an economically struggling nation to turn its land into cash.

In practice, ZEDEs have become libertarian experiments in autonomous authority. *Le Monde* described Próspera as 'a utopia for those advocating the abolition of the state'.[5] The charter city's founder, Erick Brimen, once made a tongue-in-cheek remark in an interview about 'libertopia'.[6]

The Próspera ZEDE was established in 2017 by a US company based in Delaware, Honduras Próspera LLC,[7] with $120 million from investment fund Pronomos Capital, founded by Patri Friedman, and including investment from Thiel, Altman and Andreessen. This was in the midst of President Juan Orlando Hernández's oppressive post-coup government, before he was extradited to the US and sentenced to forty-five years in prison for drug trafficking.[8] Today, around 2,000 people are residents and e-residents in this tropical paradise that also happens to be a low-tax, business-friendly oasis.

Próspera's civil law and regulatory structure is protected by its own local rules, a right granted by the Honduran constitution with the 2013 ZEDE amendment. This gives this special economic zone a significant degree of freedom to adopt its own tax and legal regimes, which generally follow US common law. The government of Honduras retains authority over certain crimes, like drug and human trafficking, money laundering, child exploitation and pornography, terrorism and organised crime; the ZEDE's private police force is tasked with maintaining law and order within Próspera in keeping with national law, but uniquely has authority over its own civil and commercial codes.[9]

The local rules that have been drawn up were devised to encourage investment and stimulate growth, including regulations around research and development of medical therapies: 'A handful of businesses operating in Próspera ZEDE on the Próspera Platform are running clinical trials,' explains the Próspera website. Elsewhere, in the Próspera community pages, the authors describe a 'flexible . . . regulatory environment designed to be one of the most innovation-friendly in the world.' The territory also allows 'expedited approval processes compared to traditional FDA timelines' by avoiding 'burdensome compliance requirements.'[10] As a result, Próspera has become something of a medical tourism hotspot.[11] Here, the motto is 'life is good and death is bad,' and 'technology is the force for civilizational progress.'[12]

Próspera isn't just an oddity in the geopolitical landscape, bounded by the Honduran constitution, but is part of a larger, snowballing plan that will make it one physical node in a longevity Network State.

The term 'Network State' was coined by Balaji Srinivasan, cryptocurrency entrepreneur, longevity enthusiast, and former general partner at Marc Andreessen's powerful Silicon Valley firm Andreessen Horowitz. His idea operates on the assumption that online communities are more powerful than their equivalent offline. As I've found in my own research, and as other internet researchers since the 1990s – like Barry Wellman, M. Lynne Markus and Nancy Baym, to name a few – have also observed, this can be a fair assumption: communities of practice – i.e. people who gather around psychological personal identities like fan cultures, political philosophies, and lifestyle tribes – can influence their members' attitudes and actions.

Some of the groups, however, discuss things that they want to play out, as they might say, IRL, like healthcare systems that prioritise radical life extension by loosening regulatory oversight. Yet conversations about disrupting the healthcare system only become truly meaningful when there's a healthcare system to transform. To have that, one needs an actual place, in which an online community can test out its ideas.

Any moderately sized university town would make for a fine community from which to recruit your volunteers. Except for one thing: those towns themselves exist within countries whose governments impose a regulatory framework. So what is needed is something to bridge the online and the offline worlds or, as Srinivasan explained in his 2022 book, *The Network State: How to Start a New Country*, 'a highly aligned online community with a capacity for collective action that crowdfunds territory around the world and eventually gains diplomatic recognition from pre-existing states'.[13]

That last part is key. This isn't a religious cult pooling their funds to buy a compound in Texas, destined to clash with sheriffs. A Network State is a group first persuading countries to give them land within borders where those countries' laws no longer apply. And then persuading enough places to do this – a few hundred acres here, a couple of city blocks there, an island somewhere else – that when linked together (online) become important enough to be recognised as a country in their own right. Albeit one in pieces, based on a single interest, and governed, inevitably, by whatever corporate decision-making tool is fashionable at the time.

So imagine these sovereign territories were set up to pursue radically extended life. You might picture a thriving marketplace of unregulated treatments, and a robust Right to Try philosophy, in which experimentation is a personal choice. You could also assume that technology development for treatments would be prioritised. And if another one, two, or even three territories elsewhere in the world managed to gain sovereignty, the different places could build up enough of a population to rival other sovereign nations, justifying (in their minds) global recognition.

Now, one might imagine that unilateral declarations of the formation of such a Network State would be ignored at best and stomped on at worst. Even an organisation such as Walmart – which has over a billion square feet of retail space around the world, an employee base that would place it as the 132nd-largest country in the world, population-wise, sandwiched between Georgia and Eritrea, and a sophisticated governance structure – would be rapidly dissuaded by its host country or countries should one day it declare itself sovereign.[14]

But for some nations, and specifically in the developing world, an influx of foreign investment in exchange for some land carved out as part of a billionaire's pet project could help boost the economy. For example, a ZEDE.

Network State ideologues believe they are building the future, using technology to start new cities and new countries. And the immortalists' fingerprints are all over the technology necessary to build both the financial infrastructure and the methods they use to govern themselves. They need the blockchain.

Very simply put, the blockchain is a giant spreadsheet that records the creation and transfer of the ownership of a 'token'. There are now many blockchains; most are aware of the Bitcoin blockchain because of the eponymous cryptocurrency token.

When it launched in January 2009, the first buyers were technologists and 'cypherpunks', an enthusiastically libertarian
movement that champions widespread use of cryptography and
privacy-enhancing technology to promote individual autonomy
and freedom from centralised authority. These activists saw it as
a new form of currency that would disrupt the financial system,
the economy, and the entire notion of money.

The reason the blockchain is a solid libertarian prospect
is that anyone who holds a token on the technology is able
do whatever they want with it, without any regulatory body
looking over their shoulder. There is no single person that holds
a magic pen and the one-and-only ledger of accounts; anyone
who buys a token has access to the exact same ledger as everyone
else, and everyone's ledger automatically updates digitally when
anything on it changes. The blockchain is referred to as a 'trust-
less' system: everyone can see what's going on, so everyone is in
charge.

That's why Bitcoin and other cryptocurrencies are what is
known as 'decentralised': their monetary value resides in no
single place. Cryptocurrency was what PayPal had aspired to
be: a token of value that exists only in a financial system that
sits outside any government's oversight. Peter Thiel was an early
investor in crypto: he endorsed Bitcoin as a hedge against central
banks' monetary policy and bought in through Founders Fund
in 2014 – one of the first institutional investors to do so.

That year, Thiel also invested in another blockchain,
Ethereum, which proposed to innovate on a different way of
organising its security. He awarded its developer, a young programmer named Vitalik Buterin, a $100,000 Thiel Foundation
grant to drop out of college and build the infrastructure for a
new civilisation.

Buterin is a prestigious yet self-effacing figure in this story; as part of the new breed of immortalists, he is fuelled by both the practicalities and the philosophies of the digital revolution. At any time, he can be anywhere; the inventor has been a digital nomad since 2013. In June 2022, Buterin posted an insider's peek into the contents of his 40-litre backpack – effectively all his worldly possessions that have taken him across 1.5 million kilometres via 360 flights: eight T-shirts, eight pairs of underwear, eight pairs of socks, shorts that double as a swimsuit, a packable jacket, an electric razor and toothbrush, a power adaptor, a portable micro-phone, a laptop stand, one heavier pair of shoes, one lighter pair, a sweater, gloves, sweatpants, tights, and of course his laptop. In his medicine bag were 'various life-extension medicines', includ-ing vitamins, the allegedly stress-relieving supplement ashwa-gandha, and metformin. A reminder: the guy was two years shy of his thirtieth birthday.[15]

Buterin grew up in Canada in a computer scientist house-hold where his dad, Dmitry, an avowed 'rabid anarcho-capitalist', advocated the idea of a stateless society where private companies provide all services. 'I see [anarcho-capitalism] helping create a much more distributed, decentralized world, with many fewer centralized decision makers and intermediaries', Dmitry, also a celebrity in the blockchain world, told journalist Lori Brown in 2018. 'I see many of our current problems stemming from the fact that decision-makers are far removed from the local situa-tion and are not impacted by the consequences. They think of themselves as brilliant, as gods – when they really have no clue'.[16] Dmitry introduced his teenage son to Bitcoin in 2011, and the boy rapidly became obsessed; the same year, Buterin co-founded *Bitcoin Magazine*, today the 'oldest and most established source of news, information and expert commentary' on the crypto-currency.[17] He served as its lead writer until 2014, while develop-ing his own crypto technology; in 2013, when he was nineteen,

Buterin improved on Bitcoin's underlying technology with his own blockchain, Ethereum.[18] He announced it in 2014; later that year, it earned him that grant from the Thiel Foundation.[19]

Now, I mentioned that cryptocurrency is only one part of the technological foundation of a Network State. The other is the blockchain, which is used to manage governance. And this is what makes Buterin's invention, Ethereum, so valuable to Network State pioneers.

Buterin recognised that there was more to the blockchain than making a (virtual) buck, so he invented a ledger that doesn't just record cryptocurrency: it also gives people the tools to declare ownership over other things, such as creative output. For example, NFTs – non-fungible tokens – are digital assets that represent ownership of an item. They're possible because of Buterin's blockchain. They are (simply) a certificate of authenticity held in digital form. They can be applied to any digital item: jpgs, gifs, virtual real estate, art or music. Other cryptocurrencies can be invented on a whim – typically called memecoins – such as Trump's $TRUMP coin, or Altman's World token, which authenticates users using an iris scan.

The technical innovation of Buterin's blockchain is the ability to hold more complex forms of data, which can in turn be the backbone for external apps, such as games or productivity tools. And, the Ethereum blockchain can also contain 'smart contracts' – agreements that automatically activate when predetermined conditions are met. These contracts eliminate the need for intermediaries, directly forging a binding contract between the parties involved. They have been used to automate financial transactions, manage supply chains, and implement voting systems. And because everyone has the record, no one – in theory – can take advantage of anyone else. These are considered such a safe form of record-keeping and governance that some sovereign nations are using them to share data and increase public transparency. All of this functionality makes the Ethereum blockchain essential to the Network State project.

I mentioned that Buterin is an immortalist. This too is Dmitry's influence: he introduced his teenage son to Ray Kurzweil when he was thirteen, giving him a copy of *The Singularity Is Near*.

'Ageing is, like, the ultimate sort of thing that is horrible and inescapable, right?' the waif-like developer explains to me in 2023 via a dodgy Zoom connection from Paris, where he was bouncing between sessions at the Ethereum Community Conference. 'It's like everything else in science: it's true right up until the decade in which it's not going to be true anymore.' His quiet voice shifts from a philosophical tone to something more pragmatic. 'Now is the time when it's not adaptive for people to believe those things anymore, because the opportunity to tackle the problem head-on actually exists.'

Over the years, Buterin has donated millions of dollars' worth of crypto to life-extension research. Since 2021, he's given $4.5 million to Eliezer Yudkowsky's MIRI, $2.4 million to the SENS Research Foundation, and $336 million to the Methuselah Foundation.[20] These sums are in crypto, so the values do change frequently. The longevity Network State is the union of his three passions: nation-state disruption, technological accelerationism, and radical life extension.

Between March and May 2023, Buterin helped to organise Zuzalu, a conference in a seaside resort in Montenegro.[21] Two hundred longtermists, EAs, singularitarians, rationalists and medical tourists were invited to attend. After paying the price of admission, they came to do cold plunges and blood draws, ate from a menu which Buterin tells me was 'based on as much of the Blueprint menu as we could practically achieve in

Montenegro on a budget of fifteen dollars a day', and imagined future societies in which life extension was available to all. It was an intellectual salon where they talked about how to govern, fund, and live long lives.

Buterin tells me they developed some foundational principles for their health oversight including how they would develop therapies, do trials, test supplements, and then deploy them in as equitable a way as possible. Their experimental longevity city was run by a 'decentralised autonomous organisation' (DAO) which ran on the Ethereum blockchain. DAOs are often used as a tool to manage and govern Network States within a system called tokenomics – the study and design of economic systems that use digital tokens and other cryptocurrencies. Tokens in DAOs are 'spent' to establish policies, incentivise participation, and manage risks. This is the governance piece of the Ethereum blockchain, and the token used at Zuzalu is called a Vita. Zuzalu's governance was organised by 'VitaDAO'.

Founded in 2021, VitaDAO is an online community of people from around the world who have gathered together to fund longevity, and want to support projects – financially or ideologically – that push forward new health innovations. 'VitaDAO's bottom line is to slow, stop, and even reverse aging, which includes the extension of healthspan as well as lifespan', their website explains. Their view is that ageing is a disease, and that death should be optional.[22] According to the VitaDAO website, as of March 2025 the organisation has funded twenty-four projects, including Zuzalu, with grants of between $200,000 and $1 million, 'deployed' $4.2 million and has $6 million 'in liquid funds' ready to be distributed.[23]

Currently there are over 9,000 members of VitaDAO who communicate on Discord, the messaging service favoured by hackers and blockchain enthusiasts. There are around 1,400 people who hold tokens, which give holders voting rights on proposals asking for funding, or to make decisions about governance within the DAO and within special projects like Zuzalu. As with any cryptocurrency, the value of a Vita token fluctuates; at the time of writing, a Vita can be purchased for $1.28.[24] Each token gives its holder one vote.

The blockchain ledger is public, and the DAO ethos is transparency, so based on voting records we can see that certain accounts have more tokens than others: these include the founders of VitaDAO, Buterin, Joe Betts-LaCroix of Retro Biosciences, Srinivasan, and global pharmaceutical giant Pfizer's investment arm, Pfizer Ventures. In 2023, Pfizer Ventures added $500,000 to a $4.1 million round raising money for the initiative.

Pfizer Ventures' involvement wasn't guaranteed with their financial investment; the VitaDAO community voted on a proposal to allow them to become token holders. As an international research and development drug company, they arguably represented a conflict of interests; VitaDAO is part of a nascent movement called Decentralised Science (DeSci), which aims to source funding for research through crowdfunding, and to decentralise the results of any outcomes – including eventual treatments. 'We see our involvement with VitaDAO as both a stakeholder and potential acquirer of future IP,' stated Pfizer Ventures' proposal.[25]

But being part of the DAO means they also help direct what kind of research is done. The question many members had was to what extent this would be the case. When they invested, a Vita token was worth approximately $4. Now, Pfizer Ventures' Vita holdings are part of a special-purpose governance vehicle voted into being by the VitaDAO community, which acts as a bridge for strategic contributors – like pharmaceutical companies or funds – and allocates an amount of tokens to the group as a whole, which is also how IP ownership is distributed; any IP from projects funded by VitaDAO is held equally across the VitaDAO membership. This is why the company did not gain control of 125,000 tokens and thereby have the opportunity to cast 125,000 votes.

Buterin and many other radical life extensionists believe DeSci is the only way to get longevity treatments out the gate, fast. This community is dedicated to funding research and development in what's usually called the 'dead zone' – that stage between a scientific discovery and treatment which is de Grey's bugbear. VitaDAO and Network States have become the pathway for renegade life scientists, biotech entrepreneurs and longevity engineers. It's a way for them to advance their research because no one else will. It's their Hail Mary pass. It's the libertarian dream.

After Zuzalu ended, spin-off ZuVillages popped up in Georgia, Ghana and Argentina, with the aim of continuing to iron out the kinks of governance, and distributing grants to applicants who had had one core team member at the first Zuzalu event for a minimum of one week. More ZuVillages were launched in Thailand and Italy in 2024. A mini-state opened its doors for a few weeks in the centre of Berlin in 2024 and another in 2025. The online longevity community is looking for the next territory where they can plant a Network State.

In March 2025, *Wired* reported that several groups that represent Próspera were drafting Congressional legislation to enact what Donald Trump had promised on the campaign trail: ten 'freedom cities', which would pay taxes like the rest of the US, but would be exempt from certain federal regulations. They hope to bring a model similar to the ZEDE to 'unused' federal land – currently national parks or monuments. 'I'd like a "longevity city" where everyone and their dog is on gene therapy,' said Mac Davis, the founder of Minicircle, the company that injected Bryan Johnson with follistatin.[26]

But is any of this really going to help the rest of us live longer, healthier lives? Or are Network States extractive, neocolonial projects, as some critics have warned?[27] In Honduras, community campaigners say the Próspera project has pushed locals out. In a paper about the ZEDE and its relationship with the surrounding, predominantly indigenous population, the UN has warned that the Honduran government could be in danger of reneging on its international obligations to protect human rights.[28]

In 2022, presidential candidate Xiomara Castro ran a successful campaign on a ticket to repeal the ZEDE amendment. She took office and has taken action to abolish it. Honduras Próspera LLC, the Delaware-based company that set up the private city, sued the Honduran state for $10.8 billion. That's roughly two-thirds of the country's 2023 state budget. The result of this legal action is yet to be seen, but losing the case would bankrupt the country and plunge the people of Honduras into even deeper economic hardship.[29] The Próspera ZEDE is still operational, at the time of writing, despite its ongoing legal disputes.

It will also set precedent for future Network States: locals will have fewer rights than residents, businesses that open within their boundaries will have no comeuppance for bad behaviour, and governance will be carried out like a Silicon Valley start-up with no restrictions.

∞

It's easy to point at the people in this book and dismiss their delusions of grandeur. They believe we are on the cusp of radically extended life, and they, uniquely in history, are the ones who have the ability to build the machines that will get us there. What hubris!

But consider this: those who fundamentally believe in the immortality project – whether they want to literally live forever, are creating an algorithm that they think will usher in the next enlightenment, or are funding a movement to overthrow the nation state so they can rebuild one in which their brand of libertarianism is in charge – are enjoying unparalleled political access. 'The fact that we have Trump, and the fact that we have a unified Congress and Senate, means that this is the time to really get this stuff done,' Dylan Livingston told me.

For the immortalists of Silicon Valley, death is not an inevitability. They can live on: as kingmakers, rulers of the world, or bits of computer code that carry their essence throughout the cosmos. But radical life extension will only be accelerated and rolled out by the people who hold the purse strings. They believe in a technological system above the law of any sovereign nation, and they believe that human beings are only as complex as computer code.

These are the ideas that are laying the groundwork for an eternal forever, and it seems like we have to follow. But we do have a choice: we can choose to live right now.

Epilogue: La Mort et le Mourant

Un mourant qui comptait plus de cent ans de vie, ✗
Se plaignait à la Mort que précipitamment
Elle le contraignait de partir tout à l'heure,
Sans qu'il eût fait son testament,
Sans l'avertir au moins. Est-il juste qu'on meure
Au pied levé ? dit-il : attendez quelque peu.*

Jean de La Fontaine, *La Mort et le Mourant*
(Death and the Dying Man), 1678

My dad was a medical professional for much of his life. A physician on the front line of mortality: both bringing in and sending off were part of the job. About two years before it was his turn, he'd carried my stepmother, Judy – his wife of almost thirty years – across the precipice. He was also a devout Catholic, spending most of his life preparing for what came next. And so when it came to facing his own time, I figured he'd be prepared.

He was not. There was still so much to do, and the disease consuming him was so debilitating. Witnessing this powerful man so quickly being unable to get up from a fall, take himself

---

* 'A dying man, who had attained/ A hundred years, counting on future still,/ To Death complained,/ That he too suddenly called him away,/ Ere he had made his will,/ And given him no notice of the day./ "What! Is it fair to strike an unseen blow?/ Postpone, O Death, a little, pray!"'

to the bathroom, or (heavens) go to church was a loss of control like I have never experienced. Nor had he.

When Dad was diagnosed with cancer, he knew what was coming. He'd been exhausted for months, and going to an oncologist was more about confirming his suspicions than seeking assurance. He decided, despite the prognosis, to commence treatment, even though the rational part of him knew he would just be making himself uncomfortable for the rest of his days.

It could have been a survival instinct. It could have been hope. It could have been faith in the medical process. It could have been because he had a new grandchild and an office in total disarray. Whatever it was, he chose to fight.

For the bulk of their careers, he and my stepmother worked with people who had Hansen's disease (formerly known as leprosy), in a centre in Louisiana where even today families fear it so much that they overlook the scientific reality of vaccines and treatments, and literally dump their loved ones on the levee and flee.

For them – she, a hand therapist, and he, an immunologist and physician – the body was both a thing that could be mapped *and* a thing that would always be unknowable. They accepted death in their profession and they ultimately had to accept it for themselves. Judy's funeral was a celebration: she was moving away from pain. And after Dad's death, his best friend told me he knew his deeds would grant him access to a better, eternal life. But as a non-religious person, without a conception of the afterlife to bring me comfort, I was lost, and I felt afraid.

For several years after they died, I was obsessed with death. I immersed myself in it, to understand why people meet it with such resistance. I spoke with people who work with the dying and the dead. I attended Death Cafés, where people gather to talk about the end of life. I made radio programmes about death in the digital age. I sat with the dying and heard the stories of their lives, and what they imagined might come next. Indirectly, that grief eventually turned into this book.

What I learned speaking with so many different people is that, for many, today's rhetoric around terminal illness is deeply aggressive. We are told to 'beat' it, and 'knock it out', so we can become 'survivors'. This drive to victory has been invoked across cultures and throughout history as people grapple with the inevitability of mortality. Framing illness as a battle can give people a sense of control, and let them face the end with courage and determination. But in the end, everyone dies, right?

I'm not suggesting lying down and giving up; many stories about illness and trying to defeat death involve triumphs and setbacks, and psychologically this is necessary for the preservation of the self. It's almost like you have to suffer to be given the chance for a longer life. In this suffering, in the battles we take to the mortal field, are we not, as Lucretius says, fighting a losing war to our detriment? 'If the life that is past and gone has been pleasant to thee,' he writes, 'why dost thou not retire like a guest sated with the banquet of life, and with calm mind embrace, thou fool, a rest that knows no care?'[1]

As I grieved, I did the same thing that the immortalists are doing, but I came to a radically different conclusion. They look to technology for answers; I look to people. I interviewed countless people who were close to their ends; one woman, younger than me, had survived three recurrences of breast cancer and wouldn't survive her fourth. She used her diagnosis to tick things off her bucket list, and spread the word – as a physician herself – about what a gift it was to embrace life while we have it. I spoke with friends who work in a hospice, who said that caring for people in their final months and weeks was the best job they ever had, because the reality of inevitable death stripped away everything and exposed the person beneath. The ability to truly be – that was what life was all about.

I spoke with an author of what I call the bible of caregiving – which I still recommend to friends to this day. Hugh Marriott wrote *The Selfish Pig's Guide to Caring* after decades of caring for his wife who was slowly deteriorating as Huntington's wasted her physical self from the inside, leaving her cognitive capacities intact. When they had received the diagnosis, they'd sold everything and bought a boat to travel the world, so that when she would later be incapacitated, she would be able to recall Tahiti in her mind's eye. That book taught me that even when you feel you've had enough, you still can't imagine living life without your loved one. And you just have to forgive yourself for being so conflicted and messy and human.

These are celebrations of life, because no one is trying to run away.

∞

Less than a week after my father died, a friend introduced me to a former colleague of theirs, a human rights lawyer named Alua Arthur who runs an organisation that prepares people for the end. I used her services to set up a comprehensive power of attorney that gives me a voice if, at the end of life, I can't speak, and this has given me immeasurable comfort.

Alua is also a death doula, helping people 'cross over', as she puts it. She lives in LA, and over the years she has guided both paupers and princes in the process. She also appeared with movie star Chris Hemsworth in an episode of his TV documentary series *Limitless*, in which he tried to push his body to its limits. The episode was about pushing death away.

She was one of the people I spoke with after my dad and stepmother died, when I was angry that they had been taken away. The rally against death does not provide any relief, she told me then. 'Being so afraid to die constricts against the idea of it, and so we live from that place rather than from an ease of settling into it.'

Everyone resists the end in their own way, she explains. Some create stories that become religions. Some grasp onto logic and reason and invest in utopias. Some dissect it out of curiosity, and some turn it into profit.

But, like the dying man in de La Fontaine's fable *Le Mort et le Mourant*, death comes to us all. The only thing we can control is how we face it.

King Gilgamesh ruled the Sumerian city of Uruk sometime between 2800 and 2500 BCE. Uruk was the first metropolis in the world, and where the first example of writing was found. Conveniently, for the purposes of our story, Gilgamesh is the first person on record who tried his utmost to live forever.

By historical accounts, he wasn't a docile ruler; he took what he wanted and incited terror in his people. But none of this made him happy because he lived in fear of death. All of this is set down in the very first surviving piece of literature, the *Epic of Gilgamesh*.

The story goes that one day he pushed his tyranny too far, and his people went to the gods and prayed they'd kill him. Instead, Gilgamesh's goddess mother sent down a wild child, Enkidu, who raised a ruckus through the forests surrounding Uruk. Gilgamesh and Enkidu met in battle, which ended in a draw. The pair became inseparable friends (by some accounts, lovers) and embarked on many dangerous adventures together.

Their reputation grew on earth and in the divine realm, and they began to attract followers. One, the goddess Ishtar, tried to seduce the king, but he rejected her advances. Rebuffed, she unleashed the Bull of Heaven. Gilgamesh and Enkidu killed it, and then the gods killed Enkidu in retribution.

Gilgamesh laments for a stone tablet and a half, until it occurs to him that he – the most powerful demi-god king in the world – should be able to find the solution to death. He sets off on a quest to meet the only mortal that the gods have made immortal, a former king named Utnapishtim. Along the way, Gilgamesh faces battle after battle as he is confronted by a series of increasingly difficult and heavily symbolic challenges.

The final stop before he reaches Utnapishtim in the 'far-away' is a tavern run by a woman named Siduri. She gives the determined king one final warning of the futility of his journey:

> Gilgamesh, where are you roaming? You will never find the eternal life that you seek. When the gods created mankind, they also created death, and they held back eternal life for themselves alone. Humans are born, they live, then they die, this is the order that the gods have decreed. But until the end comes, enjoy your life, spend it in happiness, not despair . . . Love the child who holds you by the hand, and give your wife pleasure in your embrace. That is the best way for a man to live.

Gilgamesh ignores her and arrives at the far-away exhausted. He asks a weak old man where he can find Utnapishtim; of course, this mortal-looking man is the person Gilgamesh seeks. The elderly Utnapishtim gives him a simple task: to stay awake for six days and seven nights. This is all he needs to do if he wishes to be immortal. No problem, says Gilgamesh, whose eyelids are already drooping. The king sits down and is unconscious in seconds.

With his fate sealed, Gilgamesh returns home, humbled. He's been terribly homesick, and has spent the journey reminiscing about Uruk with a travelling companion on his return journey. The people, the technologies, the farms, the government, the bridges, the city walls. So many gadgets! So much civilisation! But the thing he's most proud of is the writing: note-taking, accounting, poetry, and the first ever epic.

As Gilgamesh sees Uruk come into view before him, he realises that although he will have to grapple with the finality of his own death, he will leave behind the legacy of what he has built. And thus, the moral that still lives with us today in the hearts of celebrities, politicians, influencers, and tech billionaires: what we do will outlive us and, as in the case of what Gilgamesh carves into the base of the city wall, our deeds are what will become immortal.

∞

In early 2024, Bryan Johnson held an event not far from where I live in New York City. It was at a climbing gym where my 10-year-old goes with her friends to scramble up and down the terrain and burn off some of that magical kid energy. There were around 150 people gathered there that day, to listen to Johnson's sermon and to buy his life-giving merch. 'Don't Die' was emblazoned on his T-shirt in all-caps white letters, a rallying cry to the people who, in the room, were themselves terrified of the existential end.

While people were there to climb, there was a lot more

talk than chalk. In effect, the whole day was a Q&A with Johnson, mostly about the meaning of life. More specifically, what happens once we figure out how not to die. 'What's the point of life when there's no death?' asked several people. It's to remove the desire to kill – each other, the planet – explained Johnson. We have to start with no longer committing violence on ourselves.

Like James Strole, Johnson's resistance to death and the lengths he has gone to overcome it mean that its spectre continues to loom like a punishment waiting to be delivered.

But like all the immortalists in this book, regardless of where they rest their faith, doing something – anything – is a source of comfort. Johnson's solution is just as good as a prayer or a sacrifice or a mathematical model in navigating the unknown. 'Imagine if you have access to an algorithm that gives you the best physical, mental and spiritual health of your life,' said Johnson at one point. 'In exchange for achieving this, you do what the algorithm says. When it says to eat, you eat what it says to eat. Would you say yes?'

Would you?

The immortalists want a world where technology has authority, and we no longer have autonomy. They want it to direct our data and our evolution, and are bending national systems to their will. 'We believe Artificial Intelligence is our alchemy, our Philosopher's Stone – we are literally making sand think,' said Marc Andreessen.[2]

Should we be grateful?

In March 2025, Johnson tweeted, 'Dear humanity, I am building a religion.' It's called 'Don't Die': 'It's how we transition into the era of AI and solving death,' he wrote on X. 'This is the moment, humanity is entering a period where intelligence will either ensure its own survival or engineer its extinction.

The question is no longer philosophical, it is mathematical, computational, imminent ... Name a cult that's better than Don't Die.'[3]

Not a truer word has Bryan Johnson spoken. At the time of publication, he was still alive.

Acknowledgements

As with any long-form endeavour, there is a list of people I wish to acknowledge that could stretch a mile. I will try to keep it brief; each has my heartfelt thanks. I begin with the team at Bodley Head: my editor, Alice Skinner – I am grateful for your guidance and collaboration, and (unceasing) patience. To Graeme Hall, this book's project manager, who was so gentle in the face of protracted deadlines, Patrick Taylor, who battered the typescript into something so much better, and proofreader Robert Drew. To Susie Merry, who put it in all the right hands. To my agent, Bill Hamilton at AM Heath, who continues to stick with me through thick and thin. To my podcast team who helped hone the content of the original series: Lynne M. Jones, Joanna Humphreys, Lindsey Miller, Peter Gregson and our commissioning editor for BBC Radio 4, Dan Clarke. To Lauren Mancia, who caught the first whiff of several chapters and was so kind as to not laugh. To Erin Biba, whose eagle eye has given me much comfort by checking all my facts. To Peter McManus, my producer on the episodes of the BBC Radio 4 series *The Digital Human*, which inspired many of this book's historical and philosophical dangling participles. To the people on the other side of the screen at the London Writers' Salon writer's hour sessions, and the Shut Up and Write daily meetings: I literally could not have done it without you. To Geraldine Tchang, my accountability buddy,

259

and Hannah Kirshner, my in-person writing pal. To Tory the chicken, for clucking along with us. To my mum, who was not only a forever-rock, but also my unofficial scientific advisor. To Ben for clearing my decks, getting my words out, not laughing when I said I was going to write another one, and sharing my present. And finally, to Ripley for keeping my eye on the prize: from the first draft to the future.

Notes

INTRODUCTION

1 'We dig the past – you will too', Ponce de Leon's Fountain of Youth Archaeological Park website, fountainofyouthflorida.com/exhibits/excavations-2/

2 'Timeline of significant events', fountainofyouthflorida.com/history/1927/

3 Stephen Cave, *Immortality: The Quest to Live Forever and How It Drives Civilization* (New York: Skyhorse Publishing, 2017), pp. 14, 17

4 Thomas Nagel, 'Death', in *Mortal Questions* (New York: Cambridge University Press, 1979), pp. 1–10, p. 4. Available at dbanach.com/death.htm

5 Nick Bostrom, 'The Fable of the Dragon-Tyrant', *Journal of Medical Ethics* 31 (2005), pp. 273–7, nickbostrom.com/fable/dragon.pdf

6 Meghan O'Gieblyn, 'God in the machine: My strange journey into transhumanism', *Guardian*, 18 April 2017

CHAPTER 1

1 Robin Hilton, 'Tom Waits interviews Tom Waits', NPR, 20 May 2008, npr.org/sections/allsongs/2008/05/an_interview_with_tom_waits_by.html

2 internetlivestats.com/total-number-of-websites/

3 Larry Brilliant, 'My wish: Help me stop pandemics', TED talk, February 2006, ted.com/talks/larry_brilliant_my_wish_help_me_stop_pandemics/

4 'Summary of the 2007–2008 influenza season', CDC Archive, archive.cdc.gov/www_cdc_gov/flu/pastseasons/0708season.htm

5 Deborah Fallows, 'Search engine use', Pew Research Center, 6 August 2008, pewresearch.org/internet/2008/08/06/search-engine-use/

6 Stephen Shankland, 'Google conquers 2008 search market in U.S.', CNET, 14 January 2009, cnet.com/tech/services-and-software/google-conquers-2008-search-market-in-u-s/

7 Jeremy Ginsberg et al., 'Detecting influenza epidemics using search engine query data', *Nature* 457 (2009), pp. 1012–14, doi.org/10.1038/nature07634

8 David Lazer et al., 'The parable of Google Flu: Traps in big data analysis', *Science*, 14 March 2014, gking.harvard.edu/files/gking/files/0314policyforumff.pdf

9 Roland M. Strauss and Helena Marzo-Ortega, 'Michelangelo and medicine', *Journal of the Royal Society of Medicine* 95 (2002), pp. 514–15, doi.org/10.1177/014107680209501014

10 Roger Jones, 'Leonardo da Vinci: Anatomist', *British Journal of General Practice* 62 (2012), p. 319, doi.org/10.3399/bjgp12X649241

11 Deborah J. Brown and Calvin G. Normore, 'Automata', in *Descartes and the Ontology of Everyday Life* (Oxford: Oxford University Press, 2019), pp. 63–93

12 urbandictionary.com/define.php?term=Engineer%20Syndrome

13 Seamus O'Mahony, 'Hacking death in Dublin', *Gastroenterology* 168 (2025), pp. 1–3, gastrojournal.org/article/S0016-5085(24)05479-9/fulltext

14 'The FlyWire connectome: Neuronal wiring diagram of a complete fly brain', *Nature*, 2 October 2024, nature.com/immersive/d42859-024-00053-4/index.html

15 Ted Gross, 'Information theory – a short introduction', Medium, 15 April 2022, medium.com/towards-data-science/information-theory-a-short-introduction-a37f09959a1e

16 'Elon Musk talks Twitter, Tesla and how his brain works', TED talk, April 2022, ted.com/talks/elon_musk_elon_musk_talks_twitter_tesla_and_how_his_brain_works_live_at_ted2022

17 Armchair Expert podcast, 'David Sinclair', 16 July 2020, armchairexpertpod.com/pods/david-sinclair

18 Yuancheng Ryan Lu, Xiao Tian and David A. Sinclair, 'The Information Theory of Aging', *Nature Aging* 3 (2012), pp. 1486–99, doi.org/10.1038/s43587-023-00527-6

19 sinclair.hms.harvard.edu/research

20 Carlos E. Perez, 'Deep learning and an Information Theory of Aging', Medium, 21 December 2019, medium.com/intuitionmachine/ deep-learning-and-solving-aging-21eaaa6eec8b

21 Lex Fridman podcast, '#438 – Elon Musk: Neuralink and the future of humanity', 2 August 2024, lexfridman.com/elon-musk-and-neuralink-team-transcript/

CHAPTER 2

1   Marketplace Tech podcast, 'What do billboards say about a city?', 3 July 2024, marketplace.org/episode/2024/07/03/what-do-billboards-say-about-a-city

2   'Spannr 2022 longevity funding report', spannr.com/reports/ 2022-longevity-funding

3   Jack Harley, 'An introduction to the longevity industry: What it is, growth & companies', Spannr, spannr.com/articles/ longevity-industry-introduction

4   Thomas Franck, 'Human lifespan could soon pass 100 years thanks to medical tech, says BofA', CNBC, 8 May 2019, cnbc.com/2019/05/08/ techs-next-big-disruption-could-be-delaying-death.html

5   forbes.com/profile/peter-thiel/?list=billionaires (accessed June 2025)

6   forbes.com/profile/elon-musk/?list=billionaires (accessed June 2025)

7   forbes.com/profile/marc-andreessen/?list=billionaires (accessed June 2025)

8   Chris Anderson, 'The man who makes the future: *Wired* icon Marc Andreessen', *Wired*, 24 April 2012, wired.com/2012/04/ff-andreessen/

9   retro.bio/

10 www.forbes.com/profile/jeff-bezos/ (accessed July 2025)

11 www.altoslabs.com/about (accessed July 2025)

12 Phoebe Liu, 'Sergey Brin's $2 billion quest to tackle Parkinson's, bipolar disorder and now autism', *Forbes*, 6 February 2025, forbes. com/sites/phoebeliu/2025/02/06/sergey-brins-2-billion-quest-to-tackle-parkinsons-bipolar-disorder-and-now-autism/

13 Bobbie Johnson, 'Google invests more in DNA startup linked to co-founder', *Guardian*, 19 June 2009, theguardian.com/technology/ blog/2009/jun/19/google-dna-23andme

14 forbes.com/profile/vinod-khosla/?list=billionaires (accessed June 2025)

15 genome.gov/about-genomics/educational-resources/fact-sheets/
    human-genome-project

16 gordian.bio/

17 prenuvo.com/

18 l-nutra.com/

19 sens.org/wp-content/uploads/2021/11/SRF-11.1.1-Final-Report.pdf

20 Aubrey de Grey, 'Google's Calico: The war on aging has truly
    begun', *Time*, 18 September 2013, ideas.time.com/2013/09/18/finally-
    the-war-on-aging-has-truly-begun-2/; 'Interview with Aubrey
    de Grey', Big Think, 2 October 2009, bigthink.com/videos/
    big-think-interview-with-aubrey-de-grey/

21 mfoundation.org/

22 ted.com/talks/aubrey_de_grey_a_roadmap_to_end_aging

23 Jane Wakefield, 'Google spin-off Calico to search for answers
    to ageing', BBC News, 19 September 2013, bbc.com/news/
    technology-24158924

24 Harry McCracken and Lev Grossman, 'Google vs. death', *Time*, 30
    September 2013, time.com/574/google-vs-death/

25 De Grey, 'Google's Calico'

26 Ben Popper, 'Google's project to "cure death", Calico, announces $1.5
    billion research center', The Verge, 3 September 2014, theverge.com/
    2014/9/3/6102377/google-calico-cure-death-1-5-billion-research-abbvie

27 Mark Terry, 'Google's Calico hires another big name in quest to unlock
    healthy aging', BioSpace.com, 18 August 2016, biospace.com/google-s-
    calico-hires-another-big-name-in-quest-to-unlock-healthy-aging

28 Gary Churchill, 'The Jackson Laboratory and Calico to investigate
    basic biology of aging', The Jackson Laboratory, 26 April 2016, jax.org/
    news-and-insights/2016/april/calico-jax-aging

29 quora.com/Why-doesnt-Aubrey-de-Grey-collaborate-with-Calico/
    answer/Aubrey-de-Grey

30 Antonio Regalado, 'Sam Altman invested $180 million into a company
    trying to delay death', *MIT Technology Review*, 8 March 2023,
    technologyreview.com/2023/03/08/1069523/sam-altman-investment-
    180-million-retro-biosciences-longevity-death/

31 retro.bio/

CHAPTER 3

1 healthcare.utah.edu/mens-health/conditions/erectile-dysfunction/shockwave-therapy
2 ezra.com/blueprint
3 daveasprey.com/biohacking-infographic/
4 gminsights.com/industry-analysis/biohacking-market
5 toolsoftitans.com/
6 Charlotte Alter, 'The man who thinks he can live forever', *Time*, 20 September 2023, time.com/6315607/bryan-johnsons-quest-for-immortality/
7 Bryan Johnson, 'A plan for humanity', Medium, 15 May 2018, medium.com/future-literacy/a-plan-for-humanity-2bc04088e3d4
8 quantifiedself.com/
9 Kevin Kelly, 'What is the Quantified Self?', Quantifiedself.com, 5 October 2007, web.archive.org/web/20130125202220/http://quantified-self.com/2007/10/what-is-the-quantifiable-self/
10 moma.org/artists/39155-nicholas-felton
11 'Quantize', *The Digital Human*, BBC Radio 4, first broadcast April 2014, bbc.co.uk/programmes/b041vvw2
12 feltron.com/Facebook.html
13 Jenna Wortham, 'Your life on Facebook, in total recall', *New York Times*, 15 December 2011, nytimes.com/2011/12/16/technology/facebook-brings-back-the-past-with-new-design.html
14 Sigrid Forberg, 'How to get a free Fitbit from your health insurance provider', Moneywise.com, 12 September 2022, moneywise.com/insurance/health/fitbit-health-insurance-discount
15 statista.com/outlook/hmo/digital-health/digital-fitness-well-being/fitness-trackers/worldwide
16 Jill Rutter, '"Nudge Unit"', Institute for Government, 11 March 2020, instituteforgovernment.org.uk/article/explainer/nudge-unit
17 globalwellnessinstitute.org/press-room/statistics-and-facts
18 Ione Gamble, 'My generation is obsessed with the cult of wellness – but all that striving to be your best self can be dangerous', *Guardian*, 23 June 2022, theguardian.com/commentisfree/2022/jun/23/my-generation-is-obsessed-with-the-cult-of-wellness-but-all-that-striving-to-be-your-best-self-can-be-dangerous

19 statista.com/statistics/1239806/growth-top-fitness-mobile-apps-downloads/

20 statista.com/statistics/1239716/top-fitness-and-sport-apps-by-revenue/

21 Michał Wieczorek et al., 'Healthiness as a Virtue: The Healthism of mHealth and the Challenges to Public Health', *Public Health Ethics* 16:3 (2023), pp. 219–31, doi.org/10.1093/phe/phad019

22 Melanie L. Dobson, *Health as a Virtue: Thomas Aquinas and the Practice of Habits of Health* (Cambridge: The Lutterworth Press, 2015), pp. 27, 36

23 findingaids.lib.umich.edu/catalog/umich-bhl-0046

24 Brendan Bachmann, '"The Battle Creek Diet System": A pamphlet and birth of the fake meat industry', Library of Congress blog, 19 February 2020, blogs.loc.gov/inside_adams/2020/02/battle-creek-diet-fake-meat/

25 Dani Blum and Callie Holtermann, 'The new status symbol is a full-body M.R.I.', *New York Times*, 19 September 2023, nytimes.com/2023/09/19/well/live/mri-prenuvo-full-body-scan.html

26 Saloni Dattani et al., 'Life expectancy', Our World in Data, ourworldindata.org/life-expectancy

27 'We, the American Elderly', US Department of Commerce, September 1993, www2.census.gov/library/publications/decennial/1990/we-the-americans/we-09.pdf

28 '2023 profile of older Americans', Administration for Community Living, May 2024, web.archive.org/web/20250627205717/acl.gov/sites/default/files/Profile%20of%20OA/ACL_ProfileOlder Americans2023_508.pdf

29 'Aging in America', Institute on Aging, ioaging.org/aging-in-america/

30 oldestinbritain.nfshost.com/centenarians.php/men.php

31 'Folk wisdom', *The Digital Human*, BBC Radio 4, first broadcast March 2022, bbc.com/audio/play/m001547c

32 Lea Merone et al., 'Sex inequalities in medical research: A systematic scoping review of the literature', *Women's Health Reports* 3 (2022), pp. 49–59, doi.org/10.1089/whr.2021.0083

CHAPTER 4

1 Philippe Charlier et al., 'A gold elixir of youth in the 16th century French court', *BMJ* 339 (2009), pp. 1402–3, doi.org/10.1136/bmj.b5311

2 'Nicolas Flamel: Alchemy and the legend of the philosopher's stone', Google Arts & Culture/Science Museum, London, artsandculture.google.com/story/nicolas-flamel-alchemy-and-the-legend-of-the-philosopher%E2%80%99s-stone-science-museum/jwXhqPSro7CIJQ?hl=en

3 Wendy J. Turner, 'London businessmen and alchemists: Raising money for the Hundred Years War', in L. J. Andrew Villalon and Donald J. Kagay (eds), *The Hundred Years War (Part III)* (Leiden: Brill, 2013), pp. 333–54

4 'John XXII, Pope (ca. 1244–1334)', Encyclopedia.com, encyclopedia.com/science/encyclopedias-almanacs-transcripts-and-maps/john-xxii-pope-ca-1244-1334

5 'Three-quarters of Americans take dietary supplements; most users agree they are essential to maintaining health, CRN consumer survey finds', Council for Responsible Nutrition, 5 October 2023, crnusa.org/newsroom/three-quarters-americans-take-dietary-supplements-most-users-agree-they-are-essential

6 'Food supplements consumer research', Food Standards Agency, May 2018, food.gov.uk/sites/default/files/media/document/food-supplements-consumer-research.pdf

7 Jay Lindsay, 'Inventor sets his sights on immortality', NBC News, 13 February 2005, nbcnews.com/id/wbna6959575

8 Ouarda Djaoudene et al., 'A global overview of dietary supplements: Regulation, market trends, usage during the COVID-19 pandemic, and health effects', *Nutrients* 15 (2023), doi.org/10.3390/nu15153320

9 'How to enhance humans', *Economist*, 22 March 2025, economist.com/leaders/2025/03/20/how-to-enhance-humans

10 Djaoudene et al., 'A global overview of dietary supplements'

CHAPTER 5

1 Cynthia Kenyon, 'The first long-lived mutants: Discovery of the insulin/IGF-1 pathway for ageing', *Philosophical Transactions of the Royal Society B: Biological Sciences* 366 (2011), pp. 9–16, doi.org/10.1098/rstb.2010.0276

2 Ibid.

3   K. D. Kimura et al., '*daf*-2, an insulin receptor-like gene that regulates longevity and diapause in *Caenorhabditis elegans*', *Science* 277 (1997), pp. 942–6, doi.org/10.1126/science.277.5328.942

4   Kenyon, 'The first long-lived mutants'

5   *World Report on Ageing and Health*, WHO (2015), pp. 228, 227, who.int/publications/i/item/9789241565042

6   C.-É. Brown-Séquard, 'Note on the effects produced on man by sub cutaneous injections of a liquid obtained from the testicles of animals', *Lancet* 134 (1889), pp. 105–7

7   Michael A. Kozminski and David A. Bloom, 'A brief history of rejuvenation operations', *Journal of Urology* 187 (2012), pp. 1130–4, doi.org/10.1016/j.juro.2011.10.134

8   'Brown-Séquard's experiments with organ extract set the stage for hormone therapy', Healio, 25 March 2008, healio.com/news/hematology-oncology/20120325/brown-squard-s-experiments-with-organ-extract-set-the-stage-for-hormone-therapy

9   Thierry Gillyboeuf, 'The famous doctor who inserts monkeyglands in millionaires', *Spring: The Journal of the E. E. Cummings Society* 9 (2000), pp. 44–5, faculty.gvsu.edu/websterm/cummings/issue9/Gillybo9.htm

10   news.google.com/newspapers?nid=336&dat=19310114&id=seYoAAA AIBAJ&sjid=07UDAAAAIBAJ&pg=6866,1450862&hl=en

11   Gillyboeuf, 'The famous doctor who inserts monkeyglands in millionaires'

12   Ethel Mizrahy Cuperschmid and Tarcisio Passos Ribeiro de Campos, 'Dr. Voronoff's curious glandular xeno-implants', *História, Ciências, Saúde-Manguinhos* 14 (2007), doi.org/10.1590/S0104-59702007000300004

13   'Science: No ape-child', *Time*, 14 February 1927, time.com/archive/6660277/science-no-ape-child/

14   'Blood from young animals can revitalise old ones', *Economist*, 15 July 2017, economist.com/science-and-technology/2017/07/15/blood-from-young-animals-can-revitalise-old-ones

15   Haiguang Wang et al., 'Parabiosis in mice to study tissue residency of immune cells', *Current Protocols* 2 (2022), e446, doi.org/10.1002/cpz1.446

16   D. L. Coleman, 'Effects of parabiosis of obese with diabetes and normal mice', *Diabetologia* 9 (1973), pp. 294–8, link.springer.com/article/10.1007/BF01221857

17 Irina M. Conboy et al. 'Notch-mediated restoration of regenerative potential to aged muscle', *Science* 302 (2003), pp. 1575–7, doi.org/10.1126/science.1087573

18 Felipe Sierra and Ronald A. Kohanski, 'The role of the National Institute on Aging in the development of the field of geroscience', *Cold Spring Harbor Perspectives in Medicine* 13 (2023), a041211, doi.org/10.1101/cshperspect.a041211

19 Email correspondence

20 500candles podcast, 'Staying young with the help of young blood – with Dr Jesse Karmazin of Ambrosia Plasma', 2018, soundcloud.com/500candles/ambrosia-plasma-dr-jesse-karmazin

21 Danny Fortson, 'Blood extract of teenager on sale at just £6,000 a shot', *The Times*, 20 August 2017, thetimes.com/uk/healthcare/article/blood-extract-of-teenager-on-sale-at-just-6-000-a-shot-gz3t8pqrw

22 Lilly Dancyger, 'Clinic to offer "young blood" transfusions', *Rolling Stone*, 27 September 2018, rollingstone.com/culture/culture-news/young-blood-transfusion-new-york-clinic-730099/

23 'Statement from FDA Commissioner Scott Gottlieb . . .', FDA, 19 February 2019, web.archive.org/web/20250501103159/https:/www.fda.gov/news-events/press-announcements/statement-fda-commissioner-scott-gottlieb-md-and-director-fdas-center-biologics-evaluation-and-o

CHAPTER 6

1 Dave McGinn, 'Dexa scans are helping people "gamify" their health and longevity', *Globe and Mail*, 21 June 2024, theglobeandmail.com/life/article-dexa-scan-helping-people-gamify-health-and-longevity/

2 Alex Moshakis, 'How to live forever: Meet the extreme life-extensionists', *Guardian*, 23 June 2019, theguardian.com/global/2019/jun/23/how-to-live-forever-meet-the-extreme-life-extensionists-immortal-science

3 Jeanne Marie Laskas, 'Never say die', *Cultic Studies Journal* 11 (1994), pp. 37–55, articles2.icsahome.com/articles/never-say-die-laskas-cbj-11-1-1994. First published in *GQ* in 1991

4 Eileen Barker, 'The not-so-new religious movements: Changes in "the cult scene" over the past forty years', *Temenos* 50 (2014), pp. 235–56, eprints.lse.ac.uk/60946/

5 Laskas, 'Never say die'

6 Andy Zipser, 'The rift of eternal life', *New Times*, 6 July 1995, culteducation.com/group/923-people-unlimited/7098-the-rift-of-eternal-life-.html

7 Ryan Van Velzer, 'Immortality eludes People Unlimited founder', *Arizona Republic*, 16 November 2014, eu.azcentral.com/story/news/local/scottsdale/2014/11/16/people-unlimited-scottsdale-charles-paul-brown-immortality/19152253/

8 linkedin.com/posts/jamesstrole_why-are-raadfest-and-longevity-escape-velocity-activity-7287861607840301056-0afY/; x.com/aubreydegrey/status/1867687861559988679

9 Jeremy Cohen, 'There is mind all over the body: Immortalist and transhumanist futures', PhD thesis, 2021, macsphere.mcmaster.ca/bitstream/11375/27121/2/Cohen_Jeremy_202109_phdreligiousstudies.pdf

10 Laskas, 'Never say die'

11 Gail Tabor, 'Won't go gently: Group dedicated to living forever', *Arizona Republic*, 11 August 1991, archive.seattletimes.com/archive/19910811/1299259/wont-go-gently-group-dedicated-to-living-forever

12 Rina Raphael, '"Curing" death: Inside the conference dedicated to reversing human aging', *Fast Company*, 4 October 2018, fastcompany.com/90243453/meet-the-foot-soldiers-in-the-radical-war-on-aging

13 Chris Stokel-Walker, 'How Silicon Valley billionaires claim they've discovered the secret to everlasting life', *Telegraph*, 14 October 2019, telegraph.co.uk/health-fitness/body/super-rich-investing-immortality/

14 Gina Kalsi, 'Biohacker Bryan Johnson says he is no longer injecting himself with his teenage son's blood but has a new extreme way to stay young', *Daily Mail*, 30 January 2025, dailymail.co.uk/femail/article-14342525/Biohacker-Bryan-Johnson-says-no-longer-injecting-teenage-sons-blood-new-extreme-way-stay-young.html; x.com/bryan_johnson/status/1676636370910187520

15 x.com/bryan_johnson/status/1724449140220866571

16 x.com/bryan_johnson/status/1845950287032492378

17 Lucretius, *On the Nature of Things*, trans. Cyril Bailey (Oxford: Clarendon Press, 1910). Available at oll.libertyfund.org/titles/bailey-on-the-nature-of-things

18 Stephen Cave, *Immortality: The Quest to Live Forever and How It Drives Civilization* (New York: Skyhorse Publishing, 2017), p. 17

19 jeunessima.com/raadfest-2024-celebrating-longevity-and-innovation/

CHAPTER 7

1 x.com/AlexBlechman/status/1457842724128833538

2 Peter Diamandis, 'Longevity escape velocity: Nearing immortality?', Diamandis.com, 2 May 2024, diamandis.com/blog/longevity-escape-velocity

3 Aubrey de Grey, 'Escape velocity: Why the prospect of extreme human life extension matters now', *PLoS Biology* 2 (2004), e187, doi.org/10.1371/journal.pbio.0020187

4 Gordon E. Moore, 'Cramming more components onto integrated circuits', *Electronics* 38 (April 1965), pp. 114–17, hasler.ece.gatech.edu/Published_papers/Technology_overview/gordon_moore_1965_article.pdf

5 'Our stories – Gordon Moore about Moore's Law', ASML, YouTube, 18 December 2014, youtu.be/EzyJxAP6AQo

6 Roger Cheng, 'Moore's Law is the reason your iPhone is so thin and cheap', CNET, 16 April 2015, cnet.com/tech/mobile/moores-law-is-the-reason-why-your-iphone-is-so-thin-and-cheap/

7 Richard Elkus Jr, 'U.S. competitiveness: Where do we stand? What do we do now?', Center for Strategic and International Studies, 22 June 2021, csis.org/analysis/us-competitiveness-where-do-we-stand-what-do-we-do-now

8 Audrey Woods, 'The death of Moore's Law: What it means and what might fill the gap going forward', MIT Computer Science & Artificial Intelligence Laboratory, 2021, cap.csail.mit.edu/death-moores-law-what-it-means-and-what-might-fill-gap-going-forward

9 Charles Stross, 'Tech billionaires need to stop trying to make the science fiction they grew up on real', *Scientific American*, 20 December 2023, scientificamerican.com/article/tech-billionaires-need-to-stop-trying-to-make-the-science-fiction-they-grew-up-on-real/

10 Charles Stross, 'Crib sheet: Accelerando', Charlie's Diary blog, 28 May 2013, antipope.org/charlie/blog-static/2013/05/crib-sheet-accelerando.html

11 Charles Stross, 'Accelerando', Charlie's Diary blog, 2010, antipope.org/
charlie/blog-static/fiction/accelerando/accelerando-intro.html

12 Ray Kurzweil, 'The Law of Accelerating Returns', The Kurzweil Library,
January 2001, writingsbyraykurzweil.com/the-law-of-accelerating-returns

13 Stross, 'Tech billionaires need to stop . . .'

14 Yanis Varoufakis, *Technofeudalism: What Killed Capitalism* (London:
Penguin, 2024)

15 Vernor Vinge, 'First word', *Omni* (January 1983), p. 10, joshh.ug/195/
first_word.html

16 Stross, 'Tech billionaires need to stop . . .'

CHAPTER 8

1 Spencer E. Ante, 'Prisoners await Y2K day', *Wired*, 15 October 1998,
wired.com/1998/10/prisoners-await-y2k-day/

2 'Y2K leaves airlines flying on a wing and a prayer', *Irish Times*,
22 January 1999, irishtimes.com/business/y2k-leaves-airlines-flying-
on-a-wing-and-a-pray-1.1257367

3 'Y2k family survival guide with Leonard Nimoy', Deomon Cassette,
YouTube, 30 August 2014, youtu.be/EEhEQEG43RU; Rae Alexandra,
'Watch Leonard Nimoy scare the crap out of America over the
Y2K bug', KQED, 7 December 2020, kqed.org/arts/13890009/watch-
leonard-nimoy-scare-the-crap-out-of-america-over-the-y2k-bug; Jake
Rossen, 'Advice for an apocalypse: 10 tips from Y2K survival guides', 19
December 2024, mentalfloss.com/tips-from-y2k-survival-guides

4 'Y2K', National Museum of American History, americanhistory.si.edu/
collections/object-groups/y2k

5 'Second global meeting of national Y2K coordinators begins
at headquarters', UN press release, 22 June 1999, press.un.org/
en/1999/19990622.pi1148.html

6 Cade Metz, 'In two moves, AlphaGo and Lee Sedol redefined the
future', *Wired*, 16 March 2016, wired.com/2016/03/two-moves-alphago-
lee-sedol-redefined-future/

7 Yoo Cheong-mo, 'Go master Lee says he quits unable to win over AI
Go players', Yonhap News Agency, 27 November 2019, en.yna.co.kr/
view/AEN20191127004800315

8 John Jumper et al., 'Highly accurate protein structure prediction with AlphaFold', *Nature* 596 (2021), pp. 583–9, doi.org/10.1038/s41586-021-03819-2

9 'Reflections', Sam Altman blog, 6 January 2025, blog.samaltman.com/reflections

10 Ben Goertzel, 'Artificial General Intelligence: Concept, state of the art, and future prospects', *Journal of Artificial General Intelligence* 5 (December 2014), pp. 1–48, doi.org/10.2478/jagi-2014-0001

11 'Artificial General Intelligence (AGI) market size & overview', SNS Insider, December 2023, snsinsider.com/reports/artificial-general-intelligence-market-4174

12 'The Intelligence Age', Sam Altman blog, 23 September 2024, ia.samaltman.com/

13 'Three observations', Sam Altman blog, 9 February 2025, blog.samaltman.com/three-observations

14 Andrew Anthony, '"Eugenics on steroids": The toxic and contested legacy of Oxford's Future of Humanity Institute', *Guardian*, 28 April 2024, theguardian.com/technology/2024/apr/28/nick-bostrom-controversial-future-of-humanity-institute-closure-longtermism-affective-altruism

15 Meghan O'Gieblyn, 'Ghost in the Cloud', *n+1* 28 (*Half-Life*) (Spring 2017), nplusonemag.com/issue-28/essays/ghost-in-the-cloud/

16 Meghan O'Gieblyn, *God, Human, Animal, Machine: Technology, Metaphor, and the Search for Meaning* (New York: Doubleday, 2021), p. 49

17 O'Gieblyn, 'Ghost in the Cloud'

18 Ray Kurzweil, 'The secret to living past 120 years old? Nanobots', *Wired*, 13 June 2024, wired.com/story/the-singularity-is-nearer-book-ray-kurzweil

19 'Superintelligence: Science or fiction? Elon Musk & other great minds', Future of Life Institute, YouTube, 31 January 2017, youtu.be/ho962biiZa4

20 Sigal Samuel, 'Elon Musk reveals his plan to link your brain to your smartphone', *Vox*, 17 July 2019, vox.com/future-perfect/2019/7/17/20697812/elon-musk-neuralink-ai-brain-implant-thread-robot

21 x.com/elonmusk/status/495759307346952192; 'Machine intelligence, part 1', Sam Altman blog, 25 February 2015, blog.samaltman.com/machine-intelligence-part-1; 'Baidu CEO Robin Li interviews Bill

Gates and Elon Musk at the Boao Forum, March 29 2015; Kaiser Kuo, YouTube, 31 March 2015, youtu.be/NGoZjUfOBUs?t=1176

22 Nick Bostrom, *Superintelligence: Paths, Dangers, Strategies* (Oxford: Oxford University Press, 2014), p. 26

23 Brian Howe, 'Apocalypse how? Carrboro's Phil Torres on nanobots, biotech, A.I., and other onrushing threats to our species', *Indy Week*, 9 March 2016, indyweek.com/culture/art/apocalypse-how-carrboro-s-phil-torres-nanobots-biotech-a-i-onrushing-threats-species/

24 Ibid.

25 existential-risk.com/concept

26 Nick Bostrom, 'Existential risks: Analyzing human extinction scenarios and related hazards', *Journal of Evolution and Technology* 9 (March 2002), nickbostrom.com/existential/risks.pdf

27 S. J. Beard and Émile P. Torres, 'Ripples on the great sea of life: A brief history of Existential Risk Studies', *SSRN Electronic Journal* (March 2020), doi.org/10.2139/ssrn.3730000

28 Holden Karnofsky and Alexander Berger, 'Our progress in 2021 and plans for 2022', Open Philanthropy, 12 May 2022, openphilanthropy. org/research/our-progress-in-2021-and-plans-for-2022/

29 Beard and Torres, 'Ripples on the great sea of life'

30 'Moore's Law for everything', Sam Altman blog, 16 March 2021, moores.samaltman.com/

CHAPTER 9

1 Ed Regis, *Great Mambo Chicken and the Transhuman Condition* (Reading, MA: Addison–Wesley, 1990). Available at gwern.net/doc/ transhumanism/1990-regis-greatmambochickenandthetrans humancondition.pdf

2 docs.google.com/document/d/12xxhvL34i7AcjXtJ9phwelZ7IzHZ_ xiz-8lGwpWxucI/edit?tab=t.0

3 Eliezer Yudkowsky, 'Creating Friendly AI 1.0: The analysis and design of benevolent goal architectures', The Singularity Institute (2001), intelligence.org/files/CFAI.pdf

4 readthesequences.com/

5 David Whelan, 'The Harry Potter fan fiction author who wants to make everyone a little more rational', *Vice*, 2 March 2015, vice.com/en/ article/gq84xy/theres-something-weird-happening-in-the-world-of-harry-potter-168

6 Deku-shrub, Ruby, et al., 'Rationalist movement', Less Wrong, 28 March 2024, lesswrong.com/tag/rationalist-movement

7 Tom Abate, 'Smarter than thou? Stanford conference ponders a brave new world with machines more powerful than their creators', *San Francisco Chronicle*, 12 May 2006, web.archive.org/ web/20110211154255/http://www.sfgate.com/cgi-bin/article.cgi?f=%2Fc% 2Fa%2F2006%2F05%2F12%2FBUG9IIMG1V197.DTL

8 Tom Abate, 'Public meeting will re-examine future of artificial intelligence', *SFGate*, 6 September 2007, web.archive.org/ web/20160114083206/http://www.sfgate.com/news/article/Public-meeting-will-re-examine-future-of-2504766.php

9 'Panel with Peter Thiel, Aubrey de Grey, Eliezer Yudkowsky', singularitysummit, YouTube, 21 February 2012, youtu.be/oJDlzvqrPLk

10 'The transhumanists arrive', *Forbes*, 12 October 2009, forbes. com/2009/10/12/transhumanists-machines-technology-breakthroughs-singularity-summit.html

11 Cade Metz, 'Silicon Valley's safe space', *New York Times*, 13 February 2021, nytimes.com/2021/02/13/technology/slate-star-codex-rationalists. html

12 Keach Hagey, 'How Peter Thiel's relationship with Eliezer Yudkowsky launched the AI revolution', *Wired*, 20 May 2025, wired.com/story/ book-excerpt-the-optimist-open-ai-sam-altman/

13 Émile P. Torres, 'Nick Bostrom, longtermism, and the eternal return of eugenics', Truthdig, 23 January 2023, truthdig.com/ articles/nick-bostrom-longtermism-and-the-eternal-return-of-eugenics-2/; Cade Metz, 'Silicon Valley's safe space', *New York Times*, 13 February 2021, nytimes.com/2021/02/13/technology/ slate-star-codex-rationalists.html; Scott Alexander, 'Gender imbalances are mostly not due to offensive attitudes', Slate Star Codex, 1 August 2017, slatestarcodex.com/2017/08/01/ gender-imbalances-are-mostly-not-due-to-offensive-attitudes/

14 bjr et al., 'Bayes' Theorem', Less Wrong, 19 February 2024, lesswrong. com/w/bayes-theorem

15 'Panel with Peter Thiel, Aubrey de Grey, Eliezer Yudkowsky'

16 Satoshi_Nakamoto, 'Bayesian reasoning – explained like you're five', Less Wrong, 24 July 2015, lesswrong.com/posts/x7kL42bnATuaL4hrD/ bayesian-reasoning-explained-like-you-re-five

17 S. J. Beard and Émile P. Torres, 'Ripples on the great sea of life: A brief history of Existential Risk Studies', *SSRN Electronic Journal* (March 2020), doi.org/10.2139/ssrn.3730000

18 Metz, 'Silicon Valley's safe space'

19 David Yaffe-Bellany, 'A crypto emperor's vision: No pants, his rules', *New York Times*, 14 May 2022, nytimes.com/2022/05/14/business/sam-bankman-fried-ftx-crypto.html

20 Émile P. Torres, 'Against longtermism', *Aeon*, 19 October 2021, aeon. co/essays/why-longtermism-is-the-worlds-most-dangerous-secular-credo

21 Nick Bostrom, 'Existential risks: Analyzing human extinction scenarios and related hazards', *Journal of Evolution and Technology* 9 (March 2002), nickbostrom.com/existential/risks.pdf

22 existential-risk.com/concept

23 Torres, 'Against longtermism'

24 Eric Levitz, 'Is effective altruism to blame for Sam Bankman-Fried?', *New York*, 16 November 2022, nymag.com/intelligencer/2022/11/ effective-altruism-sam-bankman-fried-sbf-ftx-crypto.html

25 Andrew Anthony, '"Eugenics on steroids": The toxic and contested legacy of Oxford's Future of Humanity Institute', *Guardian*, 28 April 2024, theguardian.com/technology/2024/apr/28/nick-bostrom-controversial-future-of-humanity-institute-closure-longtermism-affective-altruism

26 Carla Zoe Cremer, 'Democratising risk – or how EA deals with critics', Effective Altruism Forum, 28 December 2021, forum.effectivealtruism. org/posts/gx7BEkoRbctjkyTme/democratising-risk-or-how-ea-deals-with-critics-1

27 'Statement on AI risk', Center for AI Safety, 30 May 2023, safe.ai/work/ statement-on-ai-risk

28 nickbostrom.com/ethics/ai

29 Eliezer Yudkowsky, 'Pausing AI developments isn't enough. We need to shut it all down', *Time*, 29 March 2023, time.com/6266923/ ai-eliezer-yudkowsky-open-letter-not-enough/

30 Haydn Belfield, 'There's no libertarian approach to preventing the end of the world', *Vox*, 7 March 2023, vox.com/future-perfect/2023/3/7/23618766/peter-thiel-existential-risk-oxford-union-silicon-valley-technology-artficial-intelligence

31 a16z.com/the-techno-optimist-manifesto/

32 Yudkowsky, 'Pausing AI developments isn't enough'

33 a16z.com/the-techno-optimist-manifesto/

34 Yudkowsky, 'Pausing AI developments isn't enough'

CHAPTER 10

1 'Dr. Martin Luther King on health care injustice', Physicians for a National Health Program, pnhp.org/news/ dr-martin-luther-king-on-health-care-injustice/

2 Brookie Madison, 'How much do you get paid to donate plasma?', GoodRx, 16 February 2024, goodrx.com/health-topic/finance/how-much-donating-plasma-pays?srsltid=AfmBOoqWxD6b8Esyy PyIwql9IxNjUC66KnpEoUlJnUBIY9QInZMx7oqR

3 donatesperm.com/faq/

4 Robert M. Kaplan and Arnold Milstein, 'Contributions of health care to longevity: A review of 4 estimation methods', *Annals of Family Medicine* 17 (2019), pp. 267–72, doi.org/10.1370/afm.2362

5 Sara R. Collins and Avni Gupta, 'The state of health insurance coverage in the U.S.', The Commonwealth Fund, 21 November 2024, commonwealthfund.org/publications/surveys/2024/nov/ state-health-insurance-coverage-us-2024-biennial-survey

6 E. M. Crimmins, S. H. Preston and B. Cohen (eds), *National Research Council (US) Panel on Understanding Divergent Trends in Longevity in High-Income Countries*, Chapter 7, 'The role of health care' (Washington, DC: National Academies Press US, 2011)

7 Berkeley Lovelace Jr, 'Millions at risk of losing health insurance after Trump's victory', NBC News, 7 November 2024,

nbcnews.com/health/health-news/millions-risk-losing-health-insurance-trumps-victory-rcna179146

8 'Why is America's blood plasma trade rising?', Finshots, 5 September 2024, finshots.in/archive/why-is-america-us-blood-plasma-trade-rising

9 Kathryn J. Edin and H. Luke Shaefer, 'Blood plasma, sweat, and tears', *The Atlantic*, 1 September 2015, theatlantic.com/business/archive/2015/09/poor-sell-blood/403012/

10 Heather Olsen et al., 'Bearing many burdens: Source plasma donation in the U.S.', Center for Health Research and Policy, Case Western Reserve University (2018), web.archive.org/web/20200513070544/https:/chrp.org/wp-content/uploads/2019/01/PDC-presentation-web-version.pdf

11 Catherine Chanfreau-Coffinier et al., 'Projected prevalence of actionable pharmacogenetic variants and level A drugs prescribed among US veterans health administration pharmacy users', *JAMA Network Open* 2 (2019), e195345, doi.org/10.1001/jamanetworkopen.2019.5345

12 S. Dharani and R. Kamaraj, 'A review of the regulatory challenges of personalized medicine', *Cureus* 16 (2024), e67891, doi.org/10.7759/cureus.67891

13 Ruth J. F. Loos, '15 years of genome-wide association studies and no signs of slowing down', *Nature Communications* 11 (2020), 5900, doi.org/10.1038/s41467-020-19653-5

14 *World Health Statistics 2024: Monitoring Health for the SDGs, Sustainable Development Goals*, WHO (2024), iris.who.int/bitstream/handle/10665/376869/9789240094703-eng.pdf?sequence=1

15 Matej Mikulic, 'Average cost of a precision medicine treatment worldwide from 2022 to 2027, by region', Statista, 6 January 2025, statista.com/statistics/1420940/average-cost-for-precision-medicine-treatment-globally-by-region/

16 Alex Janin, 'The longevity clinic will see you now – for $100,000', *Wall Street Journal*, 10 July 2023, wsj.com/health/longevity-clinics-aging-living-longer-2b98e773

17 spannr.com/companies

18 hooke.london/membership/compare-membership

19 cliniquelaprairie.com/programs/revitalisation/revitalisation/

20 who.int/health-topics/social-determinants-of-health

21 Usama Bilal et al., 'Socioeconomic status, life expectancy and mortality in a universal healthcare setting: An individual-level analysis of >6 million Catalan residents', *Preventive Medicine* 123 (June 2019), pp. 91–4, doi.org/10.1016/j.ypmed.2019.03.005; Silvia Stringhini et al., 'Socioeconomic status and the 25 × 25 risk factors as determinants of premature mortality: A multicohort study and meta-analysis of 1·7 million men and women', *Lancet* 389 (March 2017), pp. 1229–37, doi.org/10.1016/S0140-6736(16)32380-7; *World Report on Social Determinants of Health Equity*, WHO (2025), who.int/teams/social-determinants-of-health/equity-and-health/world-report-on-social-determinants-of-health-equity

22 data.who.int/countries/694; data.who.int/countries/392

23 *World Health Statistics 2024*

24 cdc.gov/nchs/pressroom/states/georgia/ga.htm

25 'What makes a long life? Look to your ZIP Code', Robert Wood Johnson Foundation, rwjf.org/en/insights/our-research/interactives/whereyouliveaffectshowlongyoulive.html

26 Andy Dangerfield, 'Tube map used to plot Londoners' life expectancy', BBC News, 20 July 2012, bbc.com/news/uk-england-london-18917932

27 *World Health Statistics 2024*

28 'Life in a violent country can be years shorter and much less predictable – even for those not involved in conflict', Science Daily, 5 February 2023

29 *World Report on Social Determinants of Health Equity*

30 Mirza Balaj et al., 'Effects of education on adult mortality: A global systematic review and meta-analysis', *Lancet Public Health* 9 (2024), e155–65, doi.org/10.1016/S2468-2667(23)00306-7

31 Amy R. Bentley, Shawneequa Callier and Charles N. Rotimi, 'Diversity and inclusion in genomic research: Why the uneven progress?', *Journal of Community Genetics* 8 (2017), pp. 255–66, doi.org/10.1007/s12687-017-0316-6

32 Joseph Henrich, Steven J. Heine and Ara Norenzayan, 'The weirdest people in the world?', *Behavioral and Brain Sciences* 33 (2010), pp. 61–83, doi.org/10.1017/S0140525X0999152X

33 Bentley, Callier and Rotimi, 'Diversity and inclusion in genomic research'; Segun Fatumo et al., 'A roadmap to increase diversity in genomic studies', *Nature Medicine* 28 (2022), pp. 243–50, doi.org/10.1038/s41591-021-01672-4

CHAPTER II

1   Aylin Sertkaya et al., 'Costs of drug development and research and development intensity in the US, 2000–2018', *JAMA Network Open* 7 (2024), e2415445, doi.org/10.1001/jamanetworkopen.2024.15445; Gail A. Van Norman, 'Drugs, devices, and the FDA: Part 1: An overview of approval processes for drugs', *JACC: Basic to Translational Science* 1 (2016), pp. 170–9, doi.org/10.1016/j.jacbts.2016.03.002

2   'Loyal receives FDA acceptance of Reasonable Expectation of Effectiveness for senior dog lifespan extension', Business Wire, 26 February 2025, businesswire.com/news/home/20250226676005/en/Loyal-Receives-FDA-Acceptance-of-Reasonable-Expectation-of-Effectiveness-for-Senior-Dog-Lifespan-Extension

3   Josipa Majic Predin, 'The trillion-dollar quest for healthier aging: How Hevolution Foundation is reshaping longevity research', *Forbes*, 12 September 2024, forbes.com/sites/josipamajic/2024/09/12/the-trillion-dollar-quest-for-healthier-aging-how-hevolution-foundation-is-reshaping-longevity-research/

4   xprize.org/prizes/healthspan

5   'ARPA-H launches new program aimed at extending the healthspan of Americans', Advanced Research Projects Agency for Health, 19 December 2024, arpa-h.gov/news-and-events/arpa-h-launches-new-program-aimed-extending-healthspan-americans

6   Nigishi Hotta, 'A new perspective on the biguanide, metformin therapy in type 2 diabetes and lactic acidosis', *Journal of Diabetes Investigation* 10 (2019), pp. 906–8, doi.org/10.1111/jdi.13090

7   Aimin Yang et al., 'Clinical outcomes following discontinuation of metformin in patients with type 2 diabetes and advanced chronic kidney disease in Hong Kong: A territory-wide, retrospective cohort and target trial emulation study', *eClinicalMedicine* 71 (2024), 102568, doi.org/10.1016/j.eclinm.2024.102568

8   clinicaltrials.gov/study/NCT02903511?intr=metformin&cond=renal&rank=4

9   clinicaltrials.gov/study/NCT04098666?intr=metformin&cond=Alzheimer%27s%20Disease&rank=1

10 Dae Hyun Kim and Kenneth Rockwood, 'Frailty in older adults', *New England Journal of Medicine* 391 (2024), pp. 538–48, doi.org/10.1056/NEJMra2301292

11 'Aubrey de Grey: Crusader against aging', TED, ted.com/speakers/aubrey_de_grey

12 congress.gov/crs-product/R46705

13 Nate Silver and Dhrumil Mehta, 'Both Republicans and Democrats have an age problem', FiveThirtyEight, 28 April 2014, fivethirtyeight.com/features/both-republicans-and-democrats-have-an-age-problem/

14 'China overtaking US as global research leader', *Global Health Matters* 19 (January/February 2020), fic.nih.gov/News/GlobalHealthMatters/january-february-2020/Pages/china-overtaking-us-as-global-research-leader.aspx

15 'Policymakers' guide to the longevity therapeutics industry', A4LI, 6 December 2024, a4li.org/guidetherapeutics/

16 'A call for a new National Institute for Healthy Longevity and Aging Research (NIHLAR)', A4LI, 10 August 2024, a4li.org/nihlar/

17 Berkeley Lovelace Jr, 'HHS plans to shutter or downsize several health agencies, including at CDC', NBC News, 27 March 2025, nbcnews.com/health/health-news/hhs-plans-shutter-downsize-several-health-agencies-cdc-rcna198254; Sharyn Alfonsi, 'How cuts at the National Institutes of Health could impact Americans' health', CBS News, 27 April 2025, cbsnews.com/news/nih-layoffs-budget-cuts-medical-research-60-minutes/

18 'Project Confirm: Promoting the transparency of Accelerated Approval for oncology indications', FDA, 4 December 2024, fda.gov/about-fda/oncology-center-excellence/project-confirm

19 Austin B. Frakt, 'The risks and benefits of expedited drug reviews', *JAMA* 320 (2018), pp. 225–6, doi.org/10.1001/jama.2018.8262

20 'The Advanced Approval pathway for longevity medicines', A4LI, 4 April 2023, a4li.org/the-advanced-approval-pathway-for-longevity-medicines/

21 Krishnan Vengadaraga Chary, 'Expedited drug review process: Fast, but flawed', *Journal of Pharmacology and Pharmacotherapeutics* 7 (2016), pp. 57–61, doi.org/10.4103/0976-500X.184768

22 Alessandro Demaio and Robert Marshall, 'Social lobbying: A call to arms for public health', *Lancet* 391 (2018), pp. 1558–9, doi.org/10.1016/S0140-6736(18)30831-6

23 Karen Nikos-Rose, 'Industry lobbying on WHO overshadowing public health policy, researchers suggest: Corporate dollars elevate commercial interests over health expertise', UCDavis, 18 May 2022, ucdavis.edu/news/industry-lobbying-who-overshadowing-public-health-policy-researchers-suggest

24 'Health lobbying: Annual lobbying totals, 1998–2024', OpenSecrets, opensecrets.org/industries/lobbying?cycle=2024&ind=H

25 'Labour came into power promising to clean up politics ... that promise now faces its first test ...', EveryDoctor, everydoctor.org.uk/talking-points/revealed-labour-ministers-took-donations-from-big-pharma-and-us-healthcare-lobbyists

26 Rebecca Pifer, 'Healthcare lobbying rose 70% over past two decades', Healthcare Dive, 31 October 2022, healthcaredive.com/news/healthcare-lobbying-expenditures-phrma-hospital/635337/

27 a4li.org/about/

28 Stephen Morris and Hannah Kuchler, 'Sam Altman-backed Retro Biosciences to raise $1bn for project to extend human life', *Financial Times*, 23 January 2025, ft.com/content/25a473ea-9f87-474a-8729-bc5287df853a

29 khoslaventures.com/portfolio/

30 'Introducing the Longevity Science Caucus', A4LI, 27 February 2023, a4li.org/introduction-t-the-longevity-science-caucus/

31 'Most liberal House members', *National Journal*, 23 February 2012, web.archive.org/web/20120314222109/http://www.nationaljournal.com/pictures-video/most-liberal-house-members-pictures-20120223

32 'Bilirakis, Tonko, Crenshaw, Khanna, Peters and Liccardo celebrate re-launch of Longevity Science Caucus', Gus Bilirakis press release, 25 April 2025, bilirakis.house.gov/media/press-releases/bilirakis-tonko-crenshaw-khanna-peters-and-liccardo-celebrate-re-launch

33 Jacqueline DiChiara, 'ICD-1 to ICD-11 timeline highlights healthcare's evolution', TechTarget, 11 August 2015, techtarget.com/revcyclemanagement/news/366600227/ICD-1-to-ICD-11-Timeline-Highlights-Healthcares-Evolution

34 'Gender incongruence and transgender health in the ICD', WHO, who.int/standards/classifications/frequently-asked-questions/ gender-incongruence-and-transgender-health-in-the-icd

35 'A major win for transgender rights: UN health agency drops "gender identity disorder", as official diagnosis', UN News, 30 May 2019, news. un.org/en/story/2019/05/1039531

36 'Is ageing a disease?' *Lancet Healthy Longevity* 3 (2022), e448, doi.org/10.1016/S2666-7568(22)00154-4

37 Sven Bulterijs et al., 'It is time to classify biological aging as a disease', *Frontiers in Genetics* 6 (2015), doi.org/10.3389/fgene.2015.00205

38 Víctor Manuel Mendoza-Núñez and Ana Belén Mendoza-Soto, 'Is aging a disease? A critical review within the framework of ageism', *Cureus* 16 (2024), e54834, doi.org/10.7759/cureus.54834; Kiran Rabheru et al., 'How "old age" was withdrawn as a diagnosis from ICD-11', *Lancet Healthy Longevity* 3 (2022), e457–9, doi.org/10.1016/ S2666-7568(22)00102-7

CHAPTER 12

1 John Perry Barlow, 'A declaration of the independence of cyberspace', delivered in Davos, 8 February 1996. World Economic Forum, 8 February 2018, weforum.org/stories/2018/02/ a-declaration-of-the-independence-of-cyberspace/

2 Anna Wiener, 'The complicated legacy of Stewart Brand's "Whole Earth Catalog"', *New Yorker*, 16 November 2018, newyorker.com/news/letter-from-silicon-valley/the-complicated- legacy-of-stewart-brands-whole-earth-catalog

3 Jeff Goodell, 'Who is Larry Brilliant, the guru of Google?', *Rolling Stone*, 4 April 2008, rollingstone.com/culture/culture-features/ larry-brilliant-google-guru-2008-81884/

4 Patrick Hoge, 'Doctor looks to use technology to aid global health care', *SFGate*, 24 February 2006, sfgate.com/health/article/PROFILE- Larry-Brilliant-Doctor-looks-to-use-2503524.php

5 Fred Turner, 'Where the counterculture met the new economy: The WELL and the origins of virtual community', *Technology and Culture* 46 (July 2005), pp. 485–512, doi.org/10.1353/tech.2005.0154

6 Hoge, 'Doctor looks to use technology to aid global health care'

7 Anastasia Santoreneos, 'Burning Man's billionaires: Meet the world's richest festivalgoers', *Forbes*, 6 September 2023, forbes.com.au/news/billionaires/burning-man-billionaires-worlds-richest-festivalgoers/

8 'Burning Man unofficial founder dies', Burners.me, 7 February 2018, burners.me/2018/02/07/burning-man-unofficial-founder-dies/

9 Sarah Buhr, 'Elon Musk is right, Burning Man is Silicon Valley', TechCrunch, 4 September 2014, techcrunch.com/2014/09/04/elon-musk-is-right-burning-man-is-silicon-valley/

10 Pekka Himanen, *The Hacker Ethic and the Spirit of the Information Age* (New York: Random House, 2001). Excerpt available at archive.nytimes.com/www.nytimes.com/books/first/h/himanen-hacker.html

11 Steven Levy, *Hackers: Heroes of the Computer Revolution* (New York: Doubleday, 1984), p. 28

12 Jason Bobe, 'Science without scientists', DIYbio, 22 August 2008, diybio.org/2008/08/22/science-without-scientists/

13 genspace.org/

14 'In search of a new consensus: From tension to intention', Ipsos Global Trends, September 2024, resources.ipsos.com/rs/297-CXJ-795/images/Ipsos_Global_Trends_10thAnniversary_Edition.pdf

15 Jan Greene, 'EpiPen controversy reveals complexity behind drug price tags', *Annals of Emergency Medicine* 69 (2017), pp. A16–19, doi.org/10.1016/j.annemergmed.2016.10.025

16 'FDA fact sheet: Right to Try', fda.gov/media/133864/download

17 Derek Lowe, 'Federal Right to Try', *Science*, 25 May 2018, science.org/content/blog-post/federal-right-try

18 Adam Thierer, 'FDA, biohacking & the "Right to Try" for families', Technology Liberation Front, 9 May 2016, techliberation.com/2016/05/09/fda-biohacking-the-right-to-try-for-families/

19 'Overbeck Rejuvenator, Grimsby, England, 1930', Science Museum, collection.sciencemuseumgroup.org.uk/objects/co143450/overbeck-rejuvenator-grimsby-england-1930-electrotherapy-machine

20 '80 years of the federal Food, Drug, and Cosmetic Act', FDA, 11 July 2018, fda.gov/about-fda/fda-history-exhibits/80-years-federal-food-drug-and-cosmetic-act

CHAPTER 13

1 thelongevityforum.com/

2 Andrew J. Scott, Martin Ellison and David A. Sinclair, 'The economic value of targeting aging', *Nature Aging* 1 (2021), pp. 613–23, doi.org/10. 1038/s43587-021-00080-0

3 *Global Financial Stability Report, April 2012: The Quest for Lasting Stability*, IMF (2012), doi.org/10.5089/9781616352479.082

4 Kelly Ng, 'Japan population: One in 10 people now aged 80 or older', BBC News, 19 September 2023, bbc.com/news/ world-asia-66850943

5 'Japan: Selected issues', IMF Country Report No. 12/209 (August 2012), imf.org/external/pubs/ft/scr/2012/cr12209.pdf

6 'Japan govt approves record budget for aging population, defense', *The Defense Post*, 27 December 2024, thedefensepost.com/2024/12/27/ japan-record-budget-defense/

7 'Heart disease and stroke could affect at least 60% of adults in U.S. by 2050', American Heart Association, 4 June 2024, heart.org/en/ news/2024/06/04/heart-disease-and-stroke-could-affect-at-least-60-percent-of-adults-in-us-by-2050

8 Angela B. Mariotto et al., 'Medical care costs associated with cancer survivorship in the United States', *Cancer Epidemiology, Biomarkers and Prevention* 29 (2020), pp. 1304–12, doi.org/10.1158/1055-9965.EPI-19-1534

9 Emily D. Parker et al., 'Economic costs of diabetes in the U.S. in 2022', *Diabetes Care* 47 (2024), pp. 26–43, doi.org/10.2337/dci23-0085

10 Kenji Kushida, 'Japan's aging society as a technological opportunity', Carnegie Endowment for International Peace, 3 October 2024, carnegieendowment.org/research/2024/10/ japans-aging-society-as-a-technological-opportunity

11 Maitreyi Bordia Das et al., *Silver Hues: Building Age-Ready Cities*, World Bank (2022), hdl.handle.net/10986/37259

12 Max Kozlov and *Nature* magazine, 'Planned NIH cuts threaten Americans' health, senators charge in tense hearing', *Scientific American*, 11 June 2025, scientificamerican.com/article/planned-nih-cuts-threaten-americans-health-senators-charge-in-tense-hearing/

13 thelongevityforum.com/

14 'About the human rights of older persons', OHCHR, ohchr.org/en/
special-procedures/ie-older-persons/about-human-rights-older-persons

15 Adele M. Hayutin, *New Landscapes of Population Change: A
Demographic World Tour* (Stanford, CA: Hoover Institution Press,
2022), hoover.org/research/new-landscapes-population-change

16 Julia Kagan, 'Longevity risk: What it is, how it works, special
considerations', 24 June 2021, investopedia.com/terms/l/longevityrisk.asp

17 Kathyrn Armstrong, 'France pension reforms: Macron signs pension
age rise to 64 into law', BBC News, 15 April 2023, bbc.com/news/
world-europe-65279818; Nuray Bulbul and Arielle Domb, 'UK
state pension rise: Highest and lowest retirement ages compared
worldwide', *Standard*, 25 March 2025, standard.co.uk/news/uk/uk-state-
pension-rise-retirement-ages-b1055458.html; Phoebe Zhang, 'China
raises retirement age by up to 5 years amid growing pressure from
ageing population', *South China Morning Post*, 13 September 2024,
scmp.com/news/china/politics/article/3278380/china-raises-retirement-
age-5-years-amid-growing-pressure-ageing-population

18 'The French social security system III – retirement', Centre des
Liaisons Européennes et Internationales de Sécurité Sociale (2024),
cleiss.fr/docs/regimes/regime_france/an_3.html

19 Hayutin, *New Landscapes of Population Change*

20 James Chappel, *Golden Years: How Americans Invented and Reinvented
Old Age* (New York: Basic Books, 2024), p. 277

21 'The rights of older people', UK Parliament, 19 February 2025,
publications.parliament.uk/pa/cm5901/cmselect/cmwomeq/414/report.
html

22 Nick Litsardopoulos et al., 'Work and health: International
comparisons with the UK', IES (2025), employment-studies.co.uk/
system/files/resources/files/CHWL%20international%20report%
20-%20final.pdf

23 *Global Report on Ageism*, WHO (2021), iris.who.int/bitstream/
handle/10665/340208/9789240016866-eng.pdf?sequence=1

24 Ruth A. Lamont, Hannah J. Swift and Dominic Abrams, 'A review and
meta-analysis of age-based stereotype threat: Negative stereotypes, not
facts, do the damage', *Psychology and Aging* 30 (2015), pp. 180–93, doi.
org/10.1037/a0038586

25 Becca R. Levy, 'Mind matters: Cognitive and physical effects of aging self-stereotypes', *Journals of Gerontology: Series B* 58 (July 2003), pp. P203–11, doi.org/10.1093/geronb/58.4.P203

26 E-Shien Chang et al., 'Global reach of ageism on older persons' health: A systematic review', *PLoS ONE* 15 (2020), e0220857, doi.org/10.1371/journal.pone.0220857; *Global Report on Ageism*

27 Alana Officer et al., 'Ageism, healthy life expectancy and population ageing: How are they related?', *International Journal of Environmental Research and Public Health* 17 (2020), 3159, doi.org/10.3390/ijerph17093159

CHAPTER 14

1 'The Donald Trump interview – IMPAULSIVE EP. 418', IMPAULSIVE, YouTube, 13 June 2024, youtu.be/xrFdHO7FH8w?t=2992

2 Ben Jacobs, 'Peter Thiel tells Republican convention: "I am proud to be gay"', *Guardian*, 22 July 2016, theguardian.com/technology/2016/jul/21/peter-thiel-republican-national-convention-proud-to-be-gay

3 Jessica Guynn, 'Peter Thiel to tap two for Trump transition', *USA Today*, 7 December 2016, usatoday.com/story/tech/news/2016/12/07/peter-thiel-taps-silicon-valley-associates-for-donald-trump-transition/95090708/

4 Kate Brannen and Luke Hartig, 'Disrupting the White House: Peter Thiel's influence is shaping the National Security Council', Just Security, 8 February 2017, justsecurity.org/37466/disrupting-white-house-peter-thiels-influence-shaping-national-security-council/

5 'Trump administration will "take a look" at Google following Peter Thiel's treason allegations', Fox Business, 16 July 2019, foxbusiness.com/politics/trump-administration-will-take-a-look-at-google-following-peter-thiels-treason-allegations

6 x.com/realDonaldTrump/status/1151095675213553664

7 Barton Gellman, 'Peter Thiel is taking a break from democracy', *The Atlantic*, 9 November 2023, theatlantic.com/politics/archive/2023/11/peter-thiel-2024-election-politics-investing-life-views/675946/

8 Theodore Schleifer and Alyson Krueger, 'A Trump party hosted by Peter Thiel, with all of Silicon Valley', *New York Times*, 18 January 2025,

nytimes.com/2025/01/18/us/politics/trump-party-peter-thiel-zuckerberg-vance.html

9 Jeffrey M. O'Brien, 'The PayPal mafia', *Fortune*, 26 November 2007, fortune.com/article/paypal-mafia/; Daniel Roth, 'Daniel Roth and the future of money', *Wired*, 19 March 2010, wired.com/story/daniel-roth-and-the-future-of-money/

10 Peter Thiel, 'The education of a libertarian', Cato Unbound, 13 April 2009, cato-unbound.org/2009/04/13/peter-thiel/education-libertarian/

11 Gellman, 'Peter Thiel is taking a break from democracy'

12 Justin Elliott, Patricia Callahan and James Bandler, 'Lord of the Roths: How tech mogul Peter Thiel turned a retirement account for the middle class into a $5 billion tax-free piggy bank', ProPublica, 24 June 2021, propublica.org/article/lord-of-the-roths-how-tech-mogul-peter-thiel-turned-a-retirement-account-for-the-middle-class-into-a-5-billion-dollar-tax-free-piggy-bank

13 Gellman, 'Peter Thiel is taking a break from democracy'

14 Jonathan Miles, 'The billionaire king of techtopia', Details.com, September 2011, web.archive.org/web/20130305055923/https://www.details.com/culture-trends/critical-eye/201109/peter-thiel-billionaire-paypal-facebook-internet-success#ixzz1VCNep352

15 'Introducing the Seasteading Institute', 14 April 2008, web.archive.org/web/20080822052215/http://seasteading.org/stay-in-touch/press-releases/introducing-the-seasteading-institute

16 Nellie Bowles, 'Patri Friedman makes waves with "seasteading" plan', 1 June 2011, *SFGate*, sfgate.com/news/article/Patri-Friedman-makes-waves-with-seasteading-plan-2369999.php

17 simonandschuster.com/books/Seasteading/Joe-Quirk/9781451699272

18 Oliver Wainwright, 'Seasteading – a vanity project for the rich or the future of humanity?', *Guardian*, 24 June 2020, theguardian.com/environment/2020/jun/24/seasteading-a-vanity-project-for-the-rich-or-the-future-of-humanity

19 O'Brien, 'The PayPal mafia'

20 'Peter Thiel on macroeconomics and Singularity', singularitysummit, YouTube, 21 February 2012, youtu.be/KKLDevYyE9I?t=988

21 M. G. Siegler, 'Peter Thiel has new initiative to pay kids to "stop out of school"', TechCrunch, 27 September 2010, techcrunch.com/2010/09/27/peter-thiel-drop-out-of-school/

22 Yahya Abou-Ghazala et al., '"Second American Revolution": The team behind DOGE's government overhaul', CNN, 7 February 2025, edition.cnn.com/2025/02/07/politics/musk-doge-staffers-federal-government-downsizing-invs

23 Jessica Mathews, 'Peter Thiel's protégés: A common thread runs through Trump's tech team', Fortune, 21 May 2025, fortune.com/2025/05/21/peter-thiel-silicon-valley-trump-administration-elon-musk-jd-vance/

24 Jennifer Aaker and Victoria Chang, 'Obama and the power of social media and technology', Stanford Graduate School of Business, 2009, gsb.stanford.edu/faculty-research/case-studies/obama-power-social-media-technology

25 'What is Trump's AI agenda?', The Artificial Human, BBC Radio 4, first broadcast February 2025, bbc.com/audio/play/m0027k13

26 'Fight aging with a durable business – Jim O'Neill', SENS Research Foundation, YouTube, 11 February 2015, youtu.be/9Y7oazjaSyE?t=1096

27 Patrick Wingrove, Rachael Levy and Michael Erman, 'US FDA asks fired scientists to return, including some reviewing Musk's Neuralink', Reuters, 23 February 2025, reuters.com/world/us/us-fda-asks-fired-scientists-return-including-some-reviewing-musks-neuralink-2025-02-22/

28 x.com/TrumpWarRoom/status/1860127401898049820/photo/2

CHAPTER 15

1 minicircle.io/our-therapies/

2 'Received my first gene therapy: follistatin. I am now a genetically enhanced human (GEH)', Bryan Johnson, YouTube, 13 November 2023, youtu.be/GfKsKdqTyio

3 Ana Pereyra Baron, 'The ZEDEs law in Honduras: Sanctuary for exploitation, corruption, and organized crime', Latin America Working Group, lawg.org/the-zedes-law-in-honduras-sanctuary-for-exploitation-corruption-and-organized-crime/

4　web.archive.org/web/20240226172837/https://vitalia.city/

5　Angeline Montoya, 'Prospera, the eccentric private libertarian enclave in Honduras', *Le Monde*, 16 August 2024, lemonde.fr/en/international/article/2024/08/16/prospera-the-eccentric-libertarian-and-fully-private-enclave-in-honduras_6716845_4.html

6　Brian Doherty, 'How Erick Brimen helped launch a Honduran charter city', *Reason*, January 2022, reason.com/2021/12/28/how-erick-brimen-helped-launch-a-honduran-charter-city/

7　Beth Geglia and Andrea Nuila, 'A private government in Honduras moves forward', North American Congress on Latin America, 15 February 2021, web.archive.org/web/20250615024513/https://nacla.org/news/2021/02/12/private-government-honduras-zede-prospera

8　Max Matza, 'Honduras ex-president gets 45 years for drug crimes', BBC News, 26 June 2024, bbc.com/news/articles/c2ee4j1eog60

9　'Próspera FAQs', prospera.co/news/prospera-faqs, 3 September 2024

10　'Próspera FAQs', community.prospera.co/c/open-opportunities/development-and-deployment-of-next-generation-regenerative-medicine-therapies-in-prospera

11　Sadie Whitelocks, 'Inside the secret island where wealthy people go to alter their DNA', *Daily Mail*, 9 March 2025, dailymail.co.uk/health/article-14406237/secret-island-bryan-johnson-alter-dna-prospera.html; Bennett M. Sherman, 'Biohackers convene in Honduras for unregulated gene therapy trials without FDA oversight', NAD.com, 23 February 2024, nad.com/news/biohackers-convene-in-honduras-for-unregulated-gene-therapy-trials-without-fda-oversight

12　web.archive.org/web/20241227132540/https://www.vitalia.city/

13　thenetworkstate.com/the-network-state-in-one-sentence

14　T. Ozbun, 'Retail sales area of Walmart from fiscal year 2010 to 2024, by division', Statista, 14 March 2025, statista.com/statistics/241194/sales-area-of-walmart-group-stores-since-2008; corporate.walmart.com/askwalmart/how-many-people-work-at-walmart; worldometers.info/world-population/population-by-country/

15　'My 40-liter backpack travel guide', 20 June 2022, web.archive.org/web/20220621063509/https://vitalik.ca/general/2022/06/20/backpack.html

16　Lori Brown, 'Uncensored with Dmitry Buterin: A "rabid-anarcho-capitalist" and his secret sauce', LinkedIn, 6 May 2018, linkedin.com/

pulse/uncensored-dmitry-buterin-rabid-anarcho-capitalist-his-lori-brown/

17 bitcoinmagazine.com

18 'Who is Ethereum founder Vitalik Buterin?', Crypto.com, 20 September 2024, https://crypto.com/en/university/who-is-ethereum-founder-vitalik-buterin

19 Pete Rizzo, '$100k Peter Thiel Fellowship awarded to Ethereum's Vitalik Buterin', CoinDesk, 11 September 2021, coindesk.com/markets/2014/06/05/100k-peter-thiel-fellowship-awarded-to-ethereums-vitalik-buterin

20 Colm Ó Riain, 'Our all-time largest donation, and major crypto support from Vitalik Buterin', MIRI, 13 May 2021, intelligence.org/2021/05/13/two-major-donations/; 'SENS Research Foundation receives $2.4 million Ethereum donation from Vitalik Buterin', GlobeNewswire, 2 February 2018, globenewswire.com/news-release/2018/02/02/1332410/0/en/SENS-Research-Foundation-Receives-2-4-Million-Ethereum-Donation-From-Vitalik-Buterin.html; Manish Singh, 'Ethereum creator donates meme coins worth $1 billion to help India fight COVID-19', TechCrunch, 12 May 2021, techcrunch.com/2021/05/12/vitalik-buterin-donates-1-billion-worth-of-meme-coins-to-india-covid-relief-fund/

21 Jessica Hamzelou, 'Longevity enthusiasts want to create their own independent state. They're eyeing Rhode Island', MIT Technology Review, 31 May 2023, technologyreview.com/2023/05/31/1073750/new-longevity-state-rhode-island/

22 Morgan J. Weaver, 'A beginner's guide to Vitalia', VitaDAO, 22 February 2024, vitadao.com/blog-article/a-beginners-guide-to-vitalia

23 vitadao.com

24 coingecko.com/en/coins/vitadao

25 Todd White, 'VDP-54.1 expression of interest: Pfizer Ventures', VitaDAO, August 2022, gov.vitadao.com/t/vdp-54-1-expression-of-interest-pfizer-ventures/832

26 Caroline Haskins and Vittoria Elliott, '"Startup city" groups say they're meeting Trump officials to push for deregulated "freedom cities"', Wired, 7 March 2025, wired.com/story/startup-cities-donald-trump-legislation/

27 Lucas Ropek, 'Worst new trend of 2024: Techno-colonialism and the Network State movement', Gizmodo, 27 December 2024, gizmodo. com/worst-new-trend-of-2024-techno-colonialism-and-the-network-state-movement-2000525617; Gabriel Gatehouse, 'The crypto bros who dream of crowdfunding a new country', BBC News, 20 September 2024, bbc.com/news/articles/cwyl171lyewo

28 'Las ZEDE podrían suponer serios riesgos para la garantía de los derechos humanos por parte del Estado de Honduras', UN, 8 June 2021, honduras.un.org/es/130598-las-zede-podr%C3%ADan-suponer-serios-riesgos-para-la-garant%C3% ADa-de-los-derechos-humanos-por-parte

29 Guillaume Long, 'How a start-up utopia became a nightmare for Honduras', Foreign Policy, 24 January 2024, foreignpolicy. com/2024/01/24/honduras-zedes-us-prospera-world-bank-biden-castro/

EPILOGUE

1 oll.libertyfund.org/titles/bailey-on-the-nature-of-things#Lucretius_1496_238

2 a16z.com/the-techno-optimist-manifesto/

3 x.com/bryan_johnson/status/1898062895533195497

Index

About the Author

Dr Aleks Krotoski is an award-winning international broadcaster, author and academic. She has topped the ratings for the BBC and Channel 4, winning Emmy, BAFTA, Radio Academy and Royal Society awards, and has written and presented landmark technology and social science series for both radio (with BBC Radio 4's *The Artificial Human* and *The Digital Human*) and international television (with BBC World's *The Virtual Revolution*). Krotoski has held fellowships at the University of Oxford and the London School of Economics and received an honorary doctorate from the Open University. She currently teaches at NYU Tisch School of the Arts.